Hybrid Seed Production
in Vegetables:
Rationale and Methods
in Selected Crops

Hybrid Seed Production in Vegetables: Rationale and Methods in Selected Crops has been co-published simultaneously as *Journal of New Seeds*, Volume 1, Numbers 3/4 1999.

The *Journal of New Seeds* Monographic "Separates"

Below is a list of "separates," which in serials librarianship means a special issue simultaneously published as a special journal issue or double-issue *and* as a "separate" hardbound monograph. (This is a format which we also call a "DocuSerial.")

"Separates" are published because specialized libraries or professionals may wish to purchase a specific thematic issue by itself in a format which can be separately cataloged and shelved, as opposed to purchasing the journal on an on-going basis. Faculty members may also more easily consider a "separate" for classroom adoption.

"Separates" are carefully classified separately with the major book jobbers so that the journal tie-in can be noted on new book order slips to avoid duplicate purchasing.

You may wish to visit Haworth's website at . . .

http://www.haworthpressinc.com

. . . to search our online catalog for complete tables of contents of these separates and related publications.

You may also call 1-800-HAWORTH (outside US/Canada: 607-722-5857), or Fax 1-800-895-0582 (outside US/Canada: 607-771-0012), or e-mail at:

getinfo@haworthpressinc.com

Hybrid Seed Production in Vegetables: Rationale and Methods in Selected Crops, edited by Amarjit S. Basra, PhD (Vol. 1, No. 3/4, 1999). *This essential guide will help crop scientists and growers increase the quality and yield of vegetables such as cucumbers, pumpkins, squash, peppers, onions, gourds, and the fruits watermelon and winter melon. Containing suggestions and methods for overcoming male plant sterility, inbreeding, and challenges to pollination. This book will help you successfully breed hybrid plants to produce bountiful and healthy crops.*

Hybrid Seed Production in Vegetables: Rationale and Methods in Selected Crops

Amarjit S. Basra, PhD
Editor

Hybrid Seed Production in Vegetables: Rationale and Methods in Selected Crops has been co-published simultaneously as *Journal of New Seeds*, Volume 1, Numbers 3/4 1999.

CRC Press
Taylor & Francis Group
Boca Raton   London   New York

CRC Press is an imprint of the
Taylor & Francis Group, an informa business

Hybrid Seed Production in Vegetables: Rationale and Methods in Selected Crops has been co-published simultaneously as *Journal of New Seeds* ™, Volume 1, Numbers 3/4 1999.

Reprinted 2009 by CRC Press

The development, preparation, and publication of this work has been undertaken with great care. However, the publisher, employees, editors, and agents of The Haworth Press and all imprints of The Haworth Press, Inc., including The Haworth Medical Press® and Pharmaceutical Products Press®, are not responsible for any errors contained herein or for consequences that may ensue from use of materials or information contained in this work. Opinions expressed by the author(s) are not necessarily those of The Haworth Press, Inc.

Cover design by Thomas J. Mayshock Jr.

Library of Congress Cataloging-in-Publication Data

Hybrid seed production in vegetables : rationale and methods in selected crops / Amarjit S. Basra, editor.
     p. cm.
    "Co-published simultaneously as Journal of new seeds, volume 1, numbers 3/4 1999."
    Includes bibliographical references (p.).
    ISBN 1-56022-074-0 (alk. paper)–ISBN 1-56022-075-9 (alk. paper)
    1. Vegetables–Seeds. 2. Seed technology. I. Basra, Amarjit S.

SB324.75.H93 2000
635'.042–dc21   00-022367

INDEXING & ABSTRACTING

Contributions to this publication are selectively indexed or abstracted in print, electronic, online, or CD-ROM version(s) of the reference tools and information services listed below. This list is current as of the copyright date of this publication. See the end of this section for additional notes.

- *AGRICOLA Database*
- *Biology Digest*
- *CNPIEC Reference Guide: Chinese National Directory of Foreign Periodicals*
- *FINDEX www.publist.com*
- *Foods Adlibra*
- *Food Science & Technology Abstract (FSTA)*
- *Journal of Medicinal Food*
- *Seed Abstracts*
- *South African Association for Food Science & Technology (SAAFOST)*

Special Bibliographic Notes related to special journal issues (separates) and indexing/abstracting:

- indexing/abstracting services in this list will also cover material in any "separate" that is co-published simultaneously with Haworth's special thematic journal issue or DocuSerial. Indexing/abstracting usually covers material at the article/chapter level.
- monographic co-editions are intended for either non-subscribers or libraries which intend to purchase a second copy for their circulating collections.
- monographic co-editions are reported to all jobbers/wholesalers/approval plans. The source journal is listed as the "series" to assist the prevention of duplicate purchasing in the same manner utilized for books-in-series.
- to facilitate user/access services all indexing/abstracting services are encouraged to utilize the co-indexing entry note indicated at the bottom of the first page of each article/chapter/contribution.
- this is intended to assist a library user of any reference tool (whether print, electronic, online, or CD-ROM) to locate the monographic version if the library has purchased this version but not a subscription to the source journal.
- individual articles/chapters in any Haworth publication are also available through the Haworth Document Delivery Service (HDDS).

Hybrid Seed Production in Vegetables: Rationale and Methods in Selected Crops

CONTENTS

ABOUT THE EDITOR

Amarjit S. Basra, PhD, is an eminent Botanist and Associate Professor at Punjab Agricultural University, India. He is also Editor-in-Chief of Food Products Press, an imprint of The Haworth Press, Inc. Previously, he has been a visiting scientist at the Wageningen Agricultural University of the Netherlands and the University of Western Sydney Hawkesbury in Australia. Dr. Basra is a member of several professional societies including the Crop Science Society of America, American Society of Plant Physiologists, American Society for Horticultural Science, International Society for Horticultural Science, American Institute of Biological Sciences, and the Australian Society of Plant Physiologists.

The author of over 80 research and professional publications including eight books, Dr. Basra has received coveted scientific awards and honors in recognition of his original and outstanding contributions to seed and crop science. His seed research has focused on seed quality in terms of both fundamental and applied aspects. He is frequently an invited speaker and symposium organizer at major international conferences, has given lectures, seminars, and held discussions at several universities and agricultural research centers world-wide. He provides leadership in organizing and fostering cooperation in seed and crop research at the international level.

Preface

The phenomenon of heterosis, also referred to as "hybrid vigor," underlies much of the improvement in crop yields achieved in the 20th century, in a range of crops including the vegetable crops. The exploitation of heterosis in crop production is of vital importance to face the challenge of providing food and nutritional security for an ever-increasing human population. Investment in hybrid technology is of particular relevance for developing countries, where feeding the existing population in the face of decreasing amounts of agricultural land and a declining resource base, already poses formidable problems. This is a real option to meet the huge production increases that are required in the decades ahead.

The effective use of heterosis has fostered the development of worldwide seed industry. Rapid advances in plant breeding and associated seed production technologies have served to enhance the competitiveness of hybrids by increasing crop and seed yield per hectare, reducing the costs involved, and improving seed quality.

The aim of this special volume, *Hybrid Seed Production in Vegetables: Rationale and Methods in Selected Crops,* is to understand the powerful impact of heterosis and discover how advances in hybrid seed production technologies are affecting the productivity and quality of world's selected vegetable crops, e.g., cabbage, capsicum, cucurbits and onion. Contributions on the genetics and exploitation of heterosis in crops will feature regularly in the *Journal of New Seeds* and receive priority attention. My hope is that this special issue will act as a catalyst for research and development efforts in this field.

Amarjit S. Basra

[Haworth co-indexing entry note]: "Preface." Basra, Amarjit S. Co-published simultaneously in *Journal of New Seeds* (Food Products Press, an imprint of The Haworth Press, Inc.) Vol. 1, No. 3/4, 1999, pp. xiii; and: *Hybrid Seed Production in Vegetables: Rationale and Methods in Selected Crops* (ed: Amarjit S. Basra) Food Products Press, an imprint of The Haworth Press, Inc., 2000, pp. ix. Single or multiple copies of this article are available for a fee from The Haworth Document Delivery Service [1-800-342-9678, 9:00 a.m. - 5:00 p.m. (EST). E-mail address: getinfo@haworthpressinc.com].

Rationale and Methods
for Producing Hybrid Cucurbit Seed

Richard W. Robinson

SUMMARY. Significant heterosis for earliness and yield has been reported for cucurbits. Many cucurbit genes for disease resistance are dominant, and F_1 hybrids with resistance to some diseases may be developed from crosses with only one resistant parent. Hybrid cultivars are very important for cucumber, squash, melon, and watermelon. Heterosis has also been reported for *Benincasa, Lagenaria, Luffa, Momordica,* and *Trichosanthes.*

Inbred lines without serious depression of vigor have been developed for cucumber, squash, melon, and watermelon. Some cucurbit inbred lines, however, have reduced vigor and yield. Inbreeding depression is not an important factor for producing seed of most hybrid cultivars of cucurbits.

Reciprocal crosses of cucurbits are often similar, but it may be advantageous to use the line with the largest seed or the most seed production as the female parent of a hybrid. The commercial use of the F_2 generation of cucurbit hybrids has been proposed, but F_2 cultivars of cucurbits have not become important.

Cucurbit crops are insect pollinated, and the introduction of honey bee colonies into a cucurbit seed production field may be beneficial. Isolation is needed to prevent outcrossing by insects when hybrid cucurbit seed is produced by open pollination.

Hybrid cucurbit seed has been produced by:

- Hand emasculation and pollination.
- Defloration, the manual removal of male flowers from the female parent so that all of its open pollinated seed will be hybrid.

Richard W. Robinson is Professor, Department of Horticultural Sciences, New York State Agricultural Experiment Station, 314 Hedrick Hall, Geneva, NY 14456.

[Haworth co-indexing entry note]: "Rationale and Methods for Producing Hybrid Cucurbit Seed." Robinson, Richard W. Co-published simultaneously in *Journal of New Seeds* (Food Products Press, an imprint of The Haworth Press, Inc.) Vol. 1, No. 3/4, 1999, pp. 1-47; and: *Hybrid Seed Production in Vegetables: Rationale and Methods in Selected Crops* (ed: Amarjit S. Basra) Food Products Press, an imprint of The Haworth Press, Inc., 2000, pp. 1-47. Single or multiple copies of this article are available for a fee from The Haworth Document Delivery Service [1-800-342-9678, 9:00 a.m. - 5:00 p.m. (EST). E-mail address: getinfo@haworthpressinc.com].

- Gynoecy, the use of a line with only female flowers as the female parent of a hybrid. Gynoecious lines are maintained by the application of a growth regulator to induce male flower formation, permitting self or sib polliation.
- Application of the growth regulator ethephon to monoecious cucurbits to promote the development of gynoecious sex expression of the female parent of a hybrid for the period when fruit is set.
- Male sterility. Single recessive genes for male sterility have been reported for cucumber, melon, squash, and watermelon. No interacting cytoplasmic factor is known for cucurbits, and a male sterile line used as the female parent of a hybrid will segregate for the ms gene.

Interspecific hybrid cultivars of squash have been bred that are quite productive and of good quality. Tetraploid × diploid watermelons have been crossed to produce triploid hybrids that are nearly seedless. *[Article copies available for a fee from The Haworth Document Delivery Service: 1-800-342-9678. E-mail address: getinfo@haworthpressinc.com <Website: http://www.haworthpressinc.com>]*

KEYWORDS. Bottle gourd, cucumber, melon, pumpkin, squash, watermelon, winter melon, Cucurbitaceae, *Benincasa, Citrullus, Cucumis, Cucurbita, Lagenaria, Luffa, Momordica, Trichosanthes*, growth regulators, gynoecious, heterosis, inbreeding, male sterility, pollination

IMPORTANCE OF HYBRID CUCURBITS

Allard (1960) was not optimistic about the prospects for hybrid cultivars of cucurbits. He considered most cucurbit hybrids to have insufficient heterosis to justify the extra cost of producing hybrid seed, although he left open the possibility that lines with superior combining ability might be found in the future. Today, hybrid cucurbits are extremely important (Figure 1).

F_1 hybrid cultivars of cucurbits have been commercially produced since the early 1930s. It was only after World War II, however, that hybrid cucurbits became of major importance. Most of the commercial plantings of summer squash, pickling cucumbers, melons, and watermelon are now with F_1 hybrid cultivars (Hollar, 1998). Hybrids are also used, but to a less extent, for winter squash, pumpkin, and other cucurbits. Pearson (1983) reported that hybrids comprised 43% of the cucumber seed, 56% of the summer squash seed, 10% of the winter squash seed, but only about 1% of the melon seed produced in the United States in 1980. Hybrid melon seed was mostly imported from countries with lower labor costs, and only about 1000 pounds of hybrid melon seed was then produced in the U.S.

Nishi (1967) noted a tremendous increase in the use of hybrid cucurbits in Japan after World War II. The cost of hybrid seed in 1965 was about twice that for open pollinated seed of watermelon and three times more expensive for F_1 than for open pollinated cucumber seed in Japan. Nevertheless, the advantage of hybrids was so great that they became very popular. Cucumber hybrids were grown on only 18% of the farms in the chief producing districts of Japan in 1940, but 95% in 1965. Watermelon hybrids became important sooner in Japan than in many other countries, increasing from 1% in 1940 to 96% in 1965. Melon hybrids also became important, being grown on 79% of the farms in 1965 vs. none in 1940. The use of squash hybrids, however, lagged behind that of the other major cucurbit crops, being only 10% in 1965.

The use of hybrid cucurbits in Japan has since increased even more

FIGURE 1. Harvesting hybrid cucurbit seed in California. (Photo courtesy of L. A. Hollar.)

(Yamashita, 1973). Takahashi (1987) reported that 100% of the cucumbers and watermelons and 95% of the squash and melons grown there in 1983 for fresh market were F_1 hybrids. Almost all of the melons, watermelons, and cucumbers grown in Taiwan are hybrids (Yu, 1982).

INBREEDING CUCURBITS

Some researchers have found cucurbits difficult or impossible to self pollinate. Bailey (1890) reported that many of his attempts to self pollinate squash and pumpkin failed, and he concluded that some plants were self sterile. Sinnott and Durham (1922) believed that there is a high degree of self sterility in *Cucurbita pepo*. Chekalina (1975) concluded that some plants of *Cucurbita pepo* and *C. maxima* are self incompatible. Schuster, Haghdadi, and Michael (1974a) reported that self sterility in *C. pepo* increased with inbreeding. Rosa (1927) found 'Persian' and 'Honey Dew' melons difficult to self pollinate, and suggested that there was some degree of self incompatibility.

Cummings (1904), however, was able to self pollinate *Cucurbita pepo*, and Bushnell (1920) successfully made a large number of self pollinations of Hubbard squash (*C. maxima*). Rosa (1924) tested 65 melon cultivars and found them all self fertile. Many others have also self pollinated cucurbits, and it is now generally accepted that all cucurbit crops are self compatible. Occasional plants may be self sterile, for example if male sterile, but self incompatibility is not characteristic for any cultivated species of the Cucurbitaceae. This has made it possible to develop inbred lines by self pollination and use them to produce F_1 hybrid seed.

Inbreeding Depression

Several early studies of inbreeding in the Cucurbitaceae indicated that it resulted in little if any loss of yield or vigor. Cummings and Stone (1921) and Cummings and Jenkins (1928) found that the effect on yield of inbreeding *Cucurbita maxima* was negligible. Haber (1928) developed high yielding inbred lines of *C. pepo*, and Erwin and Haber (1929) concluded that inbred lines of *Cucurbita pepo*, *C. moschata*, and *C. maxima* may be developed without any serious impairment of vigor and productiveness. Doijode and Sulladmath (1983) reported variable effects of inbreeding on fruit characteristics of different lines of *C. moschata,* but generally there was no inbreeding depression. Rosa (1927) reported that melons could be inbred without loss of fruit weight, density, or volume, and Scott (1933) determined that melons could be inbred for four to seven generations without loss of vigor. Porter (1933) found that inbreeding watermelon for four to six generations did not reduce vigor.

Sinnott and Durham (1922), however, reported that many *Cucurbita pepo* inbred lines had reduced vigor. Borghi et al. (1973) confirmed that inbreeding depression may occur in *C. pepo.* Most of their inbred lines were less vigorous than hybrids, but the inbreeding depression was less than that for corn or alfalfa. Borghi and Pironi (1974) reported inbreeding depression for some, but not all *C. pepo* cultivars. Most of the 50 *C. pepo* populations they studied exhibited inbreeding depression. Schuster, Haghdadi, and Michael (1974a) found that the greatest depression in *Cucurbita pepo* due to inbreeding was for seed yield. Fruit yield and seed protein and oil content were less affected by inbreeding.

Chekalina (1975) self pollinated *Cucurbita pepo* and *C. maxima,* and concluded that most inbred plants had reduced fruit set. A number of other quantitative and qualitative traits were also decreased, and the lines differed significantly in the degree of inbreeding depression. One generation of inbreeding did not reduce fruit set or impair other traits, but inbreeding depression was noted in the second and third generations of selfing.

Rubino and Wehner (1986) concluded that there was no important depression of yield, earliness, or quality of cucumbers inbred for six generations. El-Shawaf and Baker (1981), however, estimated that there was 40% inbreeding depression for fruits on the main stem of cucumber, although there was negligible inbreeding depression for fruits on the laterals.

Based on the early studies, cucurbits have been considered to not lose vigor when inbred despite being cross pollinated species (Allard, 1960). The more recent investigations, however, have demonstrated that inbreeding depression can occur in the Cucurbitaceae. Inbreds vary for this, and some may be very productive.

There appears to be less inbreeding depression for cucurbits than for many other cross pollinated plants. Allard (1960) suggested that this may be due to only small populations of cucurbits often being grown, due to the large amount of space required for cucurbit vines. Consequently, this results in some inbreeding despite the floral mechanism favoring outcrossing.

The parents of many commercial F_1 hybrid cucurbit cultivars are not highly inbred. They are maintained not by enforced self pollination, but by natural pollination in isolation where considerable sibbing occurs and some heterozygosity is perpetuated. Often, parental lines are developed by self pollination and pedigree selection for only a few generations, and they still have some degree of heterozygosity when mass selection of bulk populations is subsequently practiced. In some cases, an open pollinated cultivar is used as the parent of a hybrid cultivar. Thus, inbreeding depression is not an important factor for producing seed of many hybrid cultivars of cucurbits.

Natural Self Pollination

A factor related to the consequences on inbreeding cucurbits is that they have a higher rate of natural self pollination than corn and many other cross pollinated crops. This may well have eliminated many deleterious recessive genes from cucurbit populations during the course of evolution.

Rosa (1927) observed differences in natural cross pollination of melon cultivars, ranging from 5.4 to 73.2%. A monoecious melon had more natural crossing than andromonoecious cultivars. Foster (1970) confirmed that monoecious melon plants have more natural crossing than andromonoecious plants.

Ivanoff (1947) found natural crossing to be quite variable for melon, ranging from 1.4 to 100% for different plants. Ganesan (1976) reported an average rate of 6.89% natural crossing for melon plants. Nugent and Hoffman (1981) reported that the natural cross pollination rate for melon ranged from none to 20%.

Whitaker and Bohn (1952) determined that the rate of natural cross pollination for melon was variable. There appeared to be more crossing with late than with early flowers. Cross pollination was more frequent in sheltered areas than in windy locations, presumably due to differences in bee activity. The amount of crossing varied from none to over 90%.

Foster (1968a) found that the proportion of hybrid melon seed produced by open pollination varied from 10.9 to 37.6% with different plant arrangements. He concluded that the most practical arrangement to produce hybrid seed by open pollination was to plant the parents in adjoining rows of the same bed. James, Massey, and Corley (1960) also found that plant arrangement influences crossing of melon. Crossing in their plots averaged only 0.62% in one year and 0.92% in another.

Jenkins (1942) obtained 64.8 to 66.5% crossing with two planting arrangements of cucumber. Wehner and Jenkins (1985) reported an average of 36% outcrossing for cucumbers in different rows, but 17% within rows. Singletary and Moore (1965) reported 37% cross pollination for watermelon.

The amount of natural cross pollination of cucurbits varies with species, genotype, insect activity, plant arrangement, season, and other factors. Under most conditions, there appears to be sufficient natural self pollination for natural selection to eliminate many deleterious genes, resulting in little inbreeding depression. Yet, genotype and environment can be manipulated to produce enough crossing to make possible efficient hybrid seed production.

Development of Inbred Lines

Pedigree selection is often used to develop inbred parents of hybrid cultivars of cucurbits. Single seed descent has seldom been used for this purpose (Tatlioglu, 1993). The backcross method has been used to add disease resistance and gynoecious sex expression to cucumber germplasm used for producing hybrid seed (Munger, 1985). Recurrent selection can be used to develop inbred lines that are improved for traits of low heritability (Lower and Edwards, 1986).

DOMINANT GENES OF IMPORTANCE FOR HYBRIDS

Dominance of genes of the Cucurbitaceae was known long before Mendel's classical research with peas. Sageret (1826) reported that melon hybrids were not always intermediate to their parents, but resembled one parent or the other for some characteristics.

Resistance to many diseases of cucurbits is dominant. This permits the development of hybrid cultivars resistant to a disease when only one parent is resistant. It also makes possible hybrids with multiple disease resistance, when a line with dominant resistance to one disease is crossed with another line dominant for resistance to other diseases. Munger (1942), for example, crossed a fusarium wilt resistant melon with another cultivar having resistance to powdery mildew to develop a hybrid with resistance to both diseases. Hybrids of lines differing for dominant disease resistance may be the very quickest way of developing multiple disease resistant cultivars.

Single dominant genes of cucumber include those for resistance to anthracnose, bacterial wilt, scab, target leaf spot, fusarium wilt, watermelon mosaic virus, and papaya ringspot virus. Resistance in watermelon to anthracnose, fusarium wilt, and zucchini yellow mosaic virus is provided by single dominant genes. In melon, dominant genes include those conferring resistance to powdery mildew, alternaria leaf blight, fusarium wilt, gummy stem blight, papaya ringspot virus, and zucchini yellow mosaic virus. Single dominant genes of squash are known for resistance to powdery mildew and zucchini yellow mosaic virus (Robinson and Decker-Walters, 1997).

Dominant genes are also known for resistance to insects. Dominance has been reported (Robinson, 1992) for aphid resistance in melon; for resistance in squash to the squash vine borer and squash bug; for red pumpkin beetle resistance in squash, melon, and bottle gourd; fruit fly resistance in squash, melon, and watermelon; cucumber beetle resistance for squash and watermelon; pickleworm resistance of cucumber; and resistance in melon to the cucurbit caterpillar.

Other dominant genes that may be desirable in a hybrid cucumber cultivar include the *De* gene for determinant plant habit, *F* for gynoecious sex expression, and *Pc* for parthenocarpic fruit development. The dominant *B* gene of squash has been of value in F_1 hybrid cultivars; their fruit become yellow early in development, have increased vitamin A content, and do not display the disfiguring symptoms of watermelon mosaic virus when infected (Shifriss, 1988). The incompletely dominant *Bu* gene for compact, "bush" plant habit has also been important for squash breeding.

Dominant resistance may be useful for hybrids even when the inheritance of resistance is complex or unknown. Foster (1967) reported that when crown blight resistant and susceptible melons were crossed, crown blight resistance was transmitted to the hybrids to a significant practical degree.

Dominance of disease resistance may be incomplete, in which case the hybrid of resistant × susceptible parents may have a lower level of disease resistance than that of the resistant parent. A cucumber hybrid cultivar heterozygous for cucumber mosaic virus resistance is not as highly resistant as its homozygous resistant parent (Munger, 1993).

Homozygous resistance in many cases is preferable to heterozygous resistance. If undesirable recessive characteristics are associated with a dominant gene for resistance, however, the hybrid of a resistant × susceptible cross may be better than its resistant parent. Cucumber hybrids heterozygous for bacterial wilt resistance do not have the late maturity and other detrimental effects associated with homozygosity for the *Bw* gene (Munger, 1978).

Dominant undesirable traits may be of importance for the development of hybrid cultivars. Nath and Dutta (1971) reported that fruit of melon hybrids would crack at maturity if one parent had that characteristic. The undesirably long fruit shape of monoecious melons is usually transmitted to their hybrids.

RECIPROCAL CROSSES

Reciprocal crosses of cucurbits are often similar, but in some cases it may be better to produce hybrid seed with one line rather than the other as the maternal parent. One of the parental lines may be a better seed producer than the other, and the parents may differ for cytoplasmically inherited differences. A tetraploid line is used as the maternal parent to produce triploid seed for seedless watermelon since the diploid parent has better pollen fertility.

Cummings (1904) reported an unusual case of reciprocal differences in fruit set for crosses of two *Cucurbita pepo* cultivars. The cross of 'Golden Custard' × 'Crookneck' was easily made, but pollinations of the reciprocal cross usually failed to set fruit.

Passmore (1934) made reciprocal crosses with *Cucurbita pepo* lines differing in seed size. The hybrid had greater seedling vigor if a line with large seed was used as the maternal parent. Hutchins and Croston (1941) generally found no significant differences in total yield of eight reciprocal crosses of *Cucurbita maxima*, but differences were noted when they compared reciprocal crosses of large and small seeded parents. Total yield was greater when the larger seeded cultivar was used as the maternal parent.

Lana (1950) found significant differences in reciprocal crosses of *C. maxima*. The most distinctive differences were for seed characters, but reciprocal hybrids also differed for vigor and yield. In general, the F_1 seed characters resembled those of the maternal parent. Korzeniewska and Niemirowicz-Szczytt (1993) reported differences in fruit yield and fruit weight for reciprocal crosses of *C. maxima*.

In many cases, there is no significant difference between reciprocal crosses of the same parents. Grebenscikov (1967, 1975) found no difference in yield components of a reciprocal cross of *Cucurbita maxima* and only small and infrequent reciprocal differences in *C. pepo* crosses. Borghi et al. (1973) and Kasrawi (1994) did not find any difference in reciprocal crosses of *C. pepo*. Hayase and Ueda (1956) found no significant differences for 15 reciprocal crosses of *C. maxima*. No differences for reciprocal crosses of cucumber were noted by El-Shawaf and Baker (1981).

HETEROSIS

It was determined more than a century ago (Munson, 1892) that there is no immediate effect on cucurbit fruit due to cross pollination. The effect of cross pollination on fruit and plant type is not expressed until the next generation, when heterosis may be expressed.

The value of hybrid cucurbits was recognized as early as 1914 when Lumsden reported that a cross of English and French melons produced a hybrid that combined desirable characteristics of each parent. In some cases, hybrids surpass each parent in earliness, yield, or other useful characteristics due to heterosis.

Cucumber

The first report of heterosis in the Cucurbitaceae was by Hayes and Jones. In 1916 they reported that F_1 cucumber hybrids frequently outyielded their higher yielding parent. The number of fruit per plant was increased more by hybrid vigor than the other characters they studied.

Hutchins (1938) noted heterosis for yield and earliness of cucumber. He suggested that it should be possible to produce hybrid cucumber seed at a price that would not be prohibitive. Heterosis of cucumber has also been reported by Peterson and DeZeeuw (1963); Singh, Gill, and Ahluwalia (1970); Pike and Mulkey (1971); Ghaderi and

Lower (1978, 1979a, 1979b); El-Shawaf and Baker (1981); Lower, Nienhuis, and Miller (1982); Rubino and Wehner (1986); and Solanki, Seth, and Lal (1987a, 1987b, 1988). Pearson (1983) concluded that adapted cucumber hybrids have 20-30% more marketable yield than open pollinated cultivars.

Hybrid cucumber seed was produced in Japan as early as 1933 (Nishi, 1955). The first American hybrid cultivar, 'Burpee Hybrid', was bred by Oved Shifriss and introduced in 1945 by the Burpee Seed Co. (Wehner and Robinson, 1991). Barnes (1966) bred cucumber hybrids that yielded better than any commercial cultivar. Hybrid cultivars of cucumber have since become very important in the US, Europe, and elsewhere (Tatlioglu, 1993).

Squash

Bushnell (1922) did not find any yield advantage of crosses of *Cucurbita maxima* inbred lines. The inbred lines, however, were all obtained by selfing the same cultivar. Heterosis is unlikely for crosses of inbreds so genetically similar.

Hutchins and Croston (1941) determined that most of the *Cucurbita maxima* hybrids they tested were superior to their parents in total yield, but there was no heterosis for earliness. They concluded that commercial production of hybrid squash seed should be economically feasible for winter squash.

Hayase and Ueda (1956) observed heterosis for yield with most of the *C. maxima* crosses tested. The F_1 hybrids were generally higher in quality than their parents. Grebenscikov (1967, 1975) reported significant heterosis for fruit yield in *Cucurbita maxima*. Korzeniewska and Niemirowicz-Szczytt (1993) investigated the combining ability and heterosis of diallele crosses of nine inbred lines of *C. maxima*. The highest heterotic effect of the 49 hybrids was for dry matter content.

Heterosis in *Cucurbita pepo* was noted by Passmore (1934). Curtis (1939) reported that a summer squash hybrid produced female blossoms 10 days earlier than either of its parents. Its yield in the first harvest of the season was more than twice that of either parent.

Buchinger (1948) confirmed that there is significant heterosis in *Cucurbita pepo*. The fruit weight produced per plant for a hybrid was more than twice as much as for either parent, and heterosis was also observed for the number of fruit and the seed yield per plant. Heterosis in *Cucurbita pepo* was also reported by Lozonov (1969); Borghi and

Pironi (1974); Schuster, Haghdadi, and Michael (1974b); Kasrawi, (1994); and Amaya and Ortega (1996). Heterosis also occurs in *Cucurbita moschata* (Doijode and Sulladmath, 1983).

It is well established that significant heterosis can occur in *Cucurbita* crosses. Hybrid summer squash cultivars have become very important because of their increased earliness and yield.

Melon

Scott (1933) reported that F_1 melons did not express heterosis for fruit weight, but were intermediate to their inbred parents. Munger (1942), however, observed that F_1 melon hybrids generally had more total and marketable yield than their parents. He also noted that fruit of the hybrids had a higher proportion of flesh and higher soluble solids content. In most cases, the hybrid produced more early fruit than either parent. He calculated that 3000 seed of good germination should provide sufficient hybrid melon plants for an acre, and no more than 10 pollinated fruit would be needed for this amount of seed. Based on the labor costs prevailing then, it was judged that only a slight increase in yield, quality, or earliness of a melon would compensate for the extra cost of producing hybrid seed.

Bohn and Davis (1957) observed that some melon hybrids exhibited heterosis for earliness to flower and for early maturity, but others did not. Lippert and Hall (1972) also observed that various melon hybrids differed for earliness. They noted that 35 of the 45 hybrids they tested had higher fruit soluble solids than either parent, and some of the hybrids were superior to their parents in netting, rind thickness, sutures, per cent flesh, fruit shape, fruit size, earliness, or yield.

Foster (1967) compared cultivars and true breeding lines of melon with different F_1 hybrids. The hybrids produced twice as much marketable fruit as the higher yielding commercial parent and had larger fruit. In general, the netting, sutures, blossom end thickness, flesh firmness, and cavity dryness characters of the fruit tended to be improved in the hybrids, as compared with the commercial parents.

Many others have also reported heterosis in melon (Kubicki, 1962; Stino, Warid, and Abdelfattah, 1963; Bhattacharya, Kato, and Jodo, 1970; Nath and Dutta, 1971; Chadha and Nandpuri, 1980; More and Seshadri, 1980; Dixit and Kalloo, 1983; Mishra and Seshadri, 1986; Seshadri, 1986; Takahashi, 1987; Randhawa and Singh, 1990; Kitroongruang, Poo-swang, and Tokumasu, 1992; Om, Oh, and Hong,

1992; and Kim, Kim, and Chung, 1996). F_1 hybrid cultivars of melon have been developed, but their use is limited by the cost of seed. (McCreight, Nerson, and Grumet, 1993) stated that F_1 hybrid melons were then of minor importance, except for the European market. Melon hybrids are now quite important in the US (Hollar, 1998), and are of considerable importance in Japan (Takahashi, 1987) and Taiwan (Yu, 1982).

Watermelon

There are some reports of significant heterosis for watermelon, but Mohr (1986) concluded that more information is needed about the yield superiority of F_1 hybrids. Dhesi, Nandpuri, and Chand (1964) observed that F_1 hybrids outyielded the higher yielding parent by as much as 72%. Singletary and Moore (1985) reported that hybrids frequently out yielded open pollinated cultivars by 40 to 60% and were more uniform. They demonstrated that hybrid watermelon seed can be produced economically, and they concluded that the increased productiveness of hybrid watermelons would easily warrant the extra cost of F_1 seed.

Hybrid cultivars of watermelon were developed in the 1930s in Japan (Takahashi, 1987) and have since become very popular in that country. Open pollinated cultivars such as 'Charleston Grey' and 'Crimson Giant' were more popular in the U.S. than hybrids for many years, but hybrid cultivars have become increasingly important there in recent years. Earliness and disease resistance are principle reasons for the popularity of hybrid watermelon cultivars (Hollar, 1998). Triploid hybrid cultivars represent only a small share of the market in many countries, but may command a premium price since their fruit have few or no seeds.

Other Cucurbits

Heterosis has been reported for *Lagenaria* (Choudhury and Singh, 1971; and Pitchaimuthu and Sirohi, 1997), *Momordica* (Lawande and Patil, 1989, 1990; Raj, Prasdana, and Peter, 1992a; Mishra et al., 1994; and Devadas and Ramadas, 1995), *Luffa* (Kadam, Desai, and Kale, 1995), and *Trichosanthes* (Raj, Prasdana, and Peter, 1992b). F_1 hybrid cultivars have been bred for bottle gourd (Choudhury and Singh, 1971), winter melon (She et al., 1996), and bitter melon (Yu, 1982).

Gynoecious sex expression in *Luffa* (Choudhury and Thakur, 1965) could be utilized in the production of hybrid seed. Dioecious species, including oyster nut, fluted pumpkin, ivy gourd, and pointed gourd offer another opportunity to produce hybrids efficiently.

F_2 HYBRIDS

Theoretically, the F_2 generation should have only 50% of the heterosis of the F_1 generation. Concurring with this, Solanki, Seth, and Lal (1987a) found significant loss of heterosis for fruit yield and other characters of cucumber in the F_2. There are other reports, however, of F_2 cucurbit populations yielding as much or nearly as well as the F_1.

Curtis (1941) observed that the F_2 generation of a *Cucurbita pepo* cross did not differ from the F_1 in early or total fruit yield. Both the F_1 and the F_2 were significantly earlier than each parent. He proposed that F_2 seed could be used for commercial planting.

Schuster, Haghdadi, and Michael (1974b) reported that hybrid vigor of *Cucurbita pepo* crosses diminished only slightly in the F_2 generation, but decreased significantly in the next generation. He suggested that F_1 seed produced by hand could be increased by open pollination to produce F_2 seed for commercial sale.

The cost of F_2 seed should be considerably less than for the F_1, since it can be produced by open pollination without the precaution against selfing or sibbing that is required for producing F_1 seed. Nevertheless, F_2 cultivars of cucurbits have not been important.

A possible disadvantage of using F_2 seed is that there may be segregation and variability. Curtis (1941) did not consider this serious, however, and pointed out that (1) variability will be minimal if parents with similar plant and fruit type are selected, (2) variation in fruit size and, to some degree, fruit shape of summer squash are determined largely by the time of harvest since it is harvested immature, and (3) some variability may be desirable, e.g., some growers may prefer to have some segregation for earliness, rather than have most plants in a field bearing fruit on the same day. He noted that 2.4% of his F_2 plants produced female blossoms two days before the first blossoms of the F_1 opened, and he attributed this to segregation for earliness. Some segregation for earliness in F_2 hybrids may be desirable, but segregation for disease resistance or fruit type is generally not acceptable.

POLLINATION BY INSECTS
IN HYBRID SEED PRODUCTION

Insect Pollinators of Cucurbits

Pollen of cultivated cucurbits is sticky and is not windborne to any extent. Natural pollination of cucurbits is usually by insects. When insect pollination is relied on to produce hybrid seed, factors influencing insect activity are important.

Michelbacher, Smith, and Hurd (1964) reported that *Peponapis* and *Xenoglossa* gourd bees are the main pollinators of *Cucurbita* in California. Other wild pollinators included bumble bees, carpenter bees, halactid bees, stingless bees, cucumber beetles, and sphinx moth.

McGregor (1976) concluded that wild bees are very efficient pollinators of *Cucurbita*, but they are frequently so limited in number or in range that they are generally of no economic importance in pollination. He considered honey bees to be the only effective pollinator that can be provided in sufficient numbers for adequate pollination of squash, cucumber, watermelon, and melon.

Cucumber plants are not often pollinated by native bees. Martin (1970) reported that cucumbers were pollinated almost entirely by honey bees. He observed that honey bees visited male and female flowers of cucumber almost exclusively for their nectar and rarely collected pollen. In contrast, both domestic and wild bees frequently gather pollen from *Cucurbita* blossoms. Honey bees collect both pollen and nectar from melon plants.

Management of Bees in Seed Production Fields

Brewer (1974) found there were sufficient natural pollinators for producing watermelon seed in Colorado. In other situations, however, the introduction of honey bee colonies into a cucurbit field can often improve fruit set. McGregor, Levin, and Foster (1965) observed a high correlation between the number of honey bee visits to melon flowers and the number of seeds produced. McGregor and Todd (1952) considered that one colony of honey bees per acre is adequate to insure thorough pollination of melon. Peto (1950) placed five hives per acre for commercial production of hybrid cucumber seed, and Frankel and Galun (1977) recommended that at least four hives per hectare should be in a cucumber seed production field. Collison and Martin (1974) suggested adding one colony of honeybees for every 50,000 cucumber plants. The optimum number of bees to introduce varies with the weather, cultivar, and growing conditions, but they recommended a bee population sufficiently large to achieve 13 bee visits per cucumber flower from noon to 1 p.m.

Bees are most active in cucurbit fields in the morning. Thus, it is advisable to wait until later in the day to apply pesticides that are lethal to bees. It is also best to select pesticides which are least harmful to bees.

Cultural practices may influence bee activity and fruit set. Collison and Martin (1973) reported that overhead irrigation significantly reduced bee activity in a cucumber field, and moisture in pollinated flowers reduced fruit set and seed production. Foster and Levin (1967) determined that bees tended to cross pollinate melons more when they were grown in high beds than when they were planted in flat beds.

Sugar-based bee attractants have been sold commercially. If successful, they could improve natural pollination and increase the production of hybrid seed (Gubin, 1945). Ambrose et al. (1995) and Schultheis et al. (1994), however, reported that these products did not increase bee activity for cucumber or watermelon.

Other crop plants or weeds may provide competition for bees that are needed to pollinate cucurbit seed production fields. Bees may prefer to pollinate other nearby plants, rather than cucurbits. McGregor (1976) reported that melon plants secrete only about one percent as much nectar as alfalfa. Collison and Martin (1970) identified 35 plant species growing near a Michigan cucumber field that attracted honey bees when the cucumbers were in bloom. They suggested mowing weeds outside of cucumber fields in order to improve bee pollination of the cucumbers.

Foster (1968a) investigated different arrangements of parental stocks in field plantings of melons to determine the proportion of hybrid seed produced by insect pollination. The percent hybrid seed produced was the greatest when the parents were grown on the same bed, either alternating plants of the two parents in the same row or planting them in adjoining rows on the same bed, and least when the parents were grown on alternate beds. In some cases, the parents of a hybrid are planted at different dates in order to ensure that pollen of the paternal parent will be available when the first pistillate flowers of the maternal parent open (Peto, 1950).

Martin (1970) observed that gynoecious cucumber plants had good fruit set when grown 25 feet away from monoecious plants, the only source of pollen. He concluded that adequate pollination could be achieved by alternating 40 feet of rows of gynoecious plants with eight feet of rows of the monoecious pollinator.

Isolation to Prevent Outcrossing in Hybrid Seed Production

There is no basis for the supposition that cucumbers and melons will cross, and there is no need to isolate a seed production field of one crop from the other. *Cucurbita maxima* and *C. pepo* do not normally cross, but some cultivars of *C. pepo* and *C. moschata* will hybridize to some degree and should be isolated. Watermelon is compatible with citron, *Citrullus lanatus* var. *citroides*, and their seed production fields should be isolated from each other.

Different cultivars of the same species should, of course, be widely separated from each other to prevent cross pollination when grown for seed. It is particularly important to isolate ornamental gourd (*Cucurbita pepo*) from any summer squash, pumpkin, or winter squash cultivar of *C. pepo*. Some ornamental gourds have a dominant gene that bestows very bitter and toxic cucurbitacin compounds to their hybrids. *Cucurbita argyrosperma* will also cross with *C. pepo* and produce a hybrid with bitter fruit through gene interaction (Borchers and Taylor, 1988), even though both parents are nonbitter.

Kanda (1984b) recommended keeping seed production fields of different cultivars of the same *Cucurbita* species at least 1500 m apart. George (1985) suggested isolation of at least 1000 m for different seed production fields and 1500 m for stock seed increases for watermelon, cucumber, melon and squash. Whitaker and Davis (1962) considered that one quarter or preferably one half of a mile was sufficient distance between seed production fields of the same cucurbit crop. Hawthorn and Pollard (1954) recommended isolation of at least one quarter of a mile for producing seed for commercial sale, depending on conditions, but a mile or more for increasing cucurbit stock seed.

It is often difficult to arrange for such distant isolation, especially when many different seed increases are made. Rhodes, Adamson, and Bridges (1987) investigated the possibility of watermelon seed production fields being closer together. They concluded that outcrossing beyond 90 feet was rare when only native bees were present. James, Massey, and Corley (1960) reported similar results for melon; they found no evidence for crossing between plants 32 feet apart. Foster and Levin (1967), however, reported that bees carried pollen across blooming plants to pollinate melon plants at least 35 feet away. Isolation of only 100 feet or less may be acceptable for producing seed for experimental use, but more distant isolation is recommended (Hawthorn and Pollard, 1954; Whitaker and Davis, 1962) for the commercial production of hybrid cucurbit seed.

Hybrid seed production by insect pollination is generally in a field isolated from other plantings of that species. If there is not sufficient space for isolation and only a relatively small amount of hybrid seed is needed, the two parents can be grown in a cage with introduced bees (McGregor and Todd, 1952). Cage seed increases may be useful for producing seed of experimental hybrids for testing, but generally are impractical for producing large seed increases for commercial sale.

Rhodes, Adamson, and Bridges (1987) determined that outcrossing between watermelon fields was reduced when *Cucumis melo* plants were grown between them. Cucurbit species that do not cross can be grown in the same seed production field, reducing the number of isolated fields required and providing a barrier against outcrossing with the same species. There is a possibility, but only slight, that more seedless fruit will be produced by interspecific pollination when two cucurbit species are grown in the same field.

HYBRID SEED PRODUCTION BY HAND POLLINATION

Methods of hand pollinating cucurbits have previously been described (Barnes, 1947; Kanda, 1984a, 1984b; Lower and Edwards, 1986; Mohr, 1986; and Whitaker and Robinson, 1986). The flower to be hand pollinated should be closed before anthesis to prevent insect pollination. If the maternal parent is andromonoecious or hermaphroditic, anthers need to be removed from the perfect flower before dehiscence.

Male flowers for producing hybrid seed of cucumber, melon, and watermelon can be closed on the day before anthesis. Alternately, they can be collected early in the morning of anthesis before they dehisce and bees arrive, or they can be harvested the previous afternoon and kept at high humidity overnight. Squash flowers open so early in the morning that it is advisable to protect the male flower from insects on the day prior to anthesis, rather than collecting them the next morning.

Munger (1942) obtained better fruit set for melon crosses made in the afternoon prior to anthesis of the pollinated flower than on the following morning. The staminate flowers were collected on the morning of anthesis and kept in a cool, moist place until used that afternoon to pollinate the emasculated floral buds of the maternal parent. Pollinating andromonoecious melons on the afternoon prior to anthesis, rather than on the next day, is more efficient in that emasculation and pollination can be done at the same time.

Pollinations are usually completed before noon, since fruit set can be reduced later in the day under field conditions. A hand pollinated flower should be bagged, covered with a gelatin capsule, or otherwise protected from insect pollination.

The production of hybrid seed by hand emasculation and pollination is limited by the high cost of labor. It is time consuming and expensive to pollinate by hand, to guard against self pollination or outcrossing by manually protecting flowers of the maternal parent from unwanted pollen, and it is particularly burdensome to excise anthers by hand if the maternal parent has perfect (bisexual) flowers.

Some watermelon plants are gynomonoecious and anthers need to be removed from the bisexual flowers if used as the maternal parent of a hybrid. Kanda (1984a) recommended using a monoecious line with only pistillate and staminate flowers as the maternal parent to avoid this problem.

Most melon cultivars are andromonoecious, with perfect and staminate flowers on the same plant. Plants with the dominant allele of the *a* locus are monoecious and have only male and female flowers, eliminating the need for costly and tedious emasculation of perfect flowers.

Monoecious melons generally have elongated fruit, poor netting, and do not "slip" (dehisce) well. These undesirable traits tend to be dominant and diminish the value of hybrids with a monoecious parent. They can be overcome to some degree by selection for the right combination of the a^+ allele and modifiers (Foster, 1970). Monoecious melons with round or nearly round fruit have been bred by More, Seshadri, and Magdum (1987) and by H. M. Munger. Acceptable monoecious forms are now being used to produce hybrid melon seed (Munger, Kyle, and Robinson, 1993).

Fruit set for hand pollinations may be less than for insect pollinations, especially for melon. Mann and Robinson (1950) concluded, from a survey of reports in the literature, that fruit set of hand pollinated melons ranged from 5 to 41%. Their own research averaged 10% fruit set for unthinned vines, but 60 to 70% for vines with all fruit removed except for one hand pollination. Wolf and Hartman (1942) increased fruit set of hand pollinated melon to 80% by pruning the vines to one main branch and two laterals and excising apical meristems, which promoted the development of perfect flowers.

Fruit set of hand pollinated melons can be aided by growth regulator treatment. Improved fruit set for hand pollinated melon flowers can be obtained by applying benzyladenine (Bowers, 1967), indoleacetic acid (Burrell and Whitaker, 1939), or aminoethoxyvinylglycine (Natti and Loy, 1978) to the pollinated flower.

Takahashi (1987) stated that female:male parent ratios of 5:1 for cucumber, 6:1 for watermelon, and 8:1 for melon are common when hybrid seed is produced by hand pollination. If the parents of a hybrid differ for seed production, the more productive line is often used as the maternal parent.

DEFLORATION TO PRODUCE HYBRID SEED

Curtis (1939) proposed producing hybrid squash seed by a method similar to the procedure used for hybrid corn, growing one row of the male parent adjoining two or more rows of the female parent and letting them cross naturally. Instead of detaselling the female parent, as for corn, he proposed that the male flower buds be removed from the squash line used as the maternal parent of the hybrid to prevent selfing or sibbing. Instead of relying on wind pollination, pollen would be brought from the male to the female squash parent by bees. After fruit setting is completed, the plants of the male lines are usually removed to prevent their seed from becoming mixed with the hybrid seed produced by the maternal parent.

It may seem impractical to manually remove each of the many male floral buds produced by a squash plant, but the labor required to remove male flower buds can be minimized by using a highly female line with relatively few male flowers as the female parent. Also, male flower buds do not need to be removed for the entire season, but only until plants of the female parent set the maximum number of fruit that will mature. Only a few mature fruit will develop on each plant.

Lozanov (1969) concluded that the cost of hybrid *Cucurbita pepo* seed produced by the defloration method was only 1.75 times that for open pollinated seed.

Hybrid seed production by defloration, the manual removal of male flowers from the naturally pollinated female parent, has been most often been used for squash. This method was commercially used to produce hybrid seed of summer squash for many years until it was largely replaced by use of the growth regulator ethephon.

Defloration was used to produce hybrid cucumber seed as early as 1933 (Nishi, 1955; Hoshino and Furuich, 1984). Edgecomb (1946) reported that only three people were needed for a two to four week period to deflorate uniform cucumber plants in order to produce hybrid seed. The female:male ratio is typically 4:1, arranged in a pattern of eight rows of the female parent and two of the male (McDonald and Copeland, 1997).

Singletary and Moore (1965) concluded that it is more economical to produce hybrid watermelon seed by defloration than by hand pollination. Peto (1950) determined that the defloration method for producing hybrid cucumber seed cost only about half as much as hand pollination. It was also more productive, yielding 8.5 fruit per plant vs. only 1.3 for hand pollination. He also found that supplementing bee pollination with hand pollination was of no benefit when the defloration method was used to produce hybrid cucumber seed. Choudhury and Singh (1971) proposed that defloration be used to facilitate the production of hybrid bottle gourd seed.

GYNOECIOUS HYBRIDS

Gynoecious cucumber plants have only female flowers. A single major dominant gene (*F*) and modifiers determine gynoecious sex expression (Kooistra, 1967). It permits hybrid seed production simply by growing the gynoecious maternal parent in the same field as the pollen-producing male parent, with pollination accomplished by bees. All seed produced by gynoecious plants should be hybrid, since they cannot self or sib pollinate. Frankel and Galun (1977) stated that the number of rows of the gynoecious parent to those of the pollinator in hybrid cucumber seed production fields may vary, but a 3:1 ratio is quite common.

Seedstocks of gynoecious lines are produced by treatment with a growth regulator that induces male flowers to develop, permitting self or sib pollination. Frankel and Galun (1977) recommended growing the homozygous gynoecious line in an isolated field for seed increase, treating one row out of every three with a growth regulator, and relying on bees for pollination.

A single dominant gene for gynoecious sex expression is known for *Cucurbita foetidissima*. Bemis et al. (1978) crossed gynoecious and monoecious forms of *C. foetidissima* to produce experimental hybrids of buffalo gourd.

Simply inherited gynoecy has not been reported for cultivated species of *Cucurbita*, but Shifriss (1987) succeeded in developing gynoecious and predominantly female lines of *C. pepo* from crosses of monoecious parents. All F_2 plants were monoecious, but there was transgressive segregation for plants more highly female than either parent. By rigid selection for femaleness in succeeding generations, he bred the predominantly female line NJ 260 (Shifriss, 1987). Crosses with NJ 260 produced only monoecious plants in early generations, but several generations of selection for a high degree of female sex expression culminated in the development of gynoecious lines. Two of these lines were treated with GA to induce male flowers, then self pollinated to produce true breeding female lines when grown in the winter greenhouse. They were predominantly female in the field in summer, producing many female and only a few male flowers. Shifriss suggested that the highly female lines resulted from recombination of many genes.

In order to synthesize female lines of squash, Shifriss (1987) recommended selecting parents that are distantly related and highly female in sex expression. He also advised growing large populations of segregating generations, and recommended an F_2 population of 5000 plants. Rigid selection for female plants can best be accomplished in an environment that promotes male flower formation, e.g., high temperature and long photoperiod.

A U.S. patent was issued to Shifriss (1987) for this method to develop female lines of squash and other plants. When hybrid squash seed is produced with female lines, he recommended that it be at moderate temperature to prevent the maternal parent from producing an unacceptable number of male flowers in response to high temperature.

Gynoecious or predominantly female *Cucurbita* hybrids can also be obtained by interspecific crosses of certain monoecious species (Robinson, Boettger, and Shail, 1978). Highly female winter squash hybrids have been bred by crossing *Cucurbita maxima* and *C. moschata* (Kanda, 1984b).

A single locus for gynoecious sex expression is known for *Cucumis melo* (Kubicki, 1962) and *Luffa acutangula* (Choudhury and Thakur, 1965). In these species, as well as in cucumber, the gene for gynoecy will interact with a gene for andromonoecious sex expression to produce hermaphroditic plants, bearing only perfect flowers.

Frankel and Galun (1977) considered that a practical method of producing hybrid melon seed would be to find genetic lines that are strictly female in certain environments, making it possible to use them as the female parent of hybrids, but produce sufficient perfect flowers in another environment to permit seed increase of the gynoecious parent by self or sib pollination. Kubicki (1966) proposed that hybrid melon seed could be produced by crossing one line homozygous for the gynoecious gene with another line heterozygous for that gene.

Gynoecious melon inbreds have been developed by Peterson, Owens, and Rowe (1983) and More, Seshadri, and Magdum (1987). Gynoecious melon hybrids have been bred, but McCreight, Nerson, and Grumet (1993) believed that they lacked quality. Gynoecy has not been widely used to produce hybrid melon seed (Munger, Kyle, and Robinson, 1993).

Gynoecious × Monoecious Hybrids

Peterson and Wiegle (1957) produced hybrid cucumber seed by crossing a line segregating for gynoecious plants with a monoecious line, after rogueing monoecious plants from the maternal parent. It was not until gynoecious inbreds not requiring rogueing were developed, however, that gynoecious cucumber hybrids became important (Peterson, 1960).

Researchers in Israel and Russia (Frankel and Galun, 1977) and Japan (Yamashita, 1973) also proposed that hybrid cucumber seed could be produced by crossing gynoecious and monoecious lines. Other systems of producing gynoecious hybrid seed have subsequently been proposed, but gynoecious × monoecious hybrids are still the most widely grown.

The first gynoecious hybrid pickling cucumber cultivar was 'Spartan Dawn', introduced in 1963. Peterson and DeZeeuw (1963) backcrossed gene *F* into the monoecious cultivar 'Wisconsin SMR 18' to produce the gynoecious line MSU 713-5, which has been used as the female parent of 'Spartan Dawn' and other cucumber hybrids.

Gynoecious hybrid cultivars produced by crossing gynoecious and monoecious lines soon became very important, particularly for mechanically harvested pickling cucumbers. Gynoecious hybrids are advantageous for mechanical harvesting, not only because of their hybrid vigor, but even more importantly because of their sex expression. Partial dominance of the gynoecious gene usually confers a high degree of female sex expression to gynoecious hybrids, and the abundance of pistillate flowers results in uniform and concentrated fruiting. Consequently, high yield is often possible in a single mechanical harvest.

Wehner and Miller (1985) compared gynoecious and monoecious slicing cucumbers having the same gene background. There was little difference in yield or quality in multiple harvests, but the gynoecious hybrids were earlier and produced a greater proportion of culls.

Many commercial seedlots of gynoecious hybrid cucumber cultivars are a blend of gynoecious and monoecious types. Peterson and DeZeeuw (1963) recommended blending in 10% of seed of a monoecious cultivar with gynoecious hybrid seed. Seed of the monoecious parent of the hybrid can be used for this purpose. Miller (1976) reported that commercial seed of gynoecious cucumber hybrids generally includes 10 to 15% of a monoecious cultivar. He found that blends with a high proportion of gynoecious seed had a greater early yield than those with more monoecious seed.

The monoecious cultivar 'Sumter' has often been used as a blender for gynoecious or predominantly female picking cucumber cultivars. Recently, seed company breeders have developed monoecious cucumber lines specifically for use as blenders, selecting for early pollen production and for fruit type similar to the hybrid that the blender will pollinate (Groff, 1993).

Blending is done to improve pollination, but it has the disadvantage of compromising the uniformity that is an important attribute of hybrids. If the gynoecious hybrid is earlier than its monoecious pollinator, then the monoecious plants may not yet have fruit when the field is mechanically harvested.

Blending is not needed if the gynoecious hybrid is parthenocarpic and does not require pollination for fruit set. Genetic parthencarpy has been bred into gynoecious hybrid cucumbers for greenhouse production.

A problem that has been encountered with many gynoecious × monoecious hybrid cucumber cultivars is that they are not dependably gynoecious. The occurrence of monoecious plants in plantings of gynoecious hybrid cultivars can have serious consequences on uniformity, earliness, and adaptation to mechanical harvesting.

Gene background is important for this to occur, and some cultivars heterozygous for the gynoecious gene are more reliably gynoecious or predominantly female than others (Kubicki, 1965). Environment is also quite important, and the proportion of gynoecious plants produced by the same seed lot of a hybrid may vary in different plantings (Lower and Edwards, 1986).

Some gynoecious inbred lines are more dependably gynoecious than others. More and Seshadri (1988) bred cucumber lines stable for gynoecious sex expression at high temperature and long photoperiod, conditions that promoted male flower formation on other lines homozygous for the gynoecious gene.

Gynoecious × Gynoecious Hybrids

Cucumber hybrids homozygous for F, the female sex expression gene, have the important advantage of being more stable for gynoecious sex expression than comparable hybrids heterozygous for that gene. They are more likely to produce entirely female plants, especially under environmental conditions or gene background that promote the development of male flowers.

Homozygosity for gynoecious sex expression is especially important for parthenocarpic greenhouse cucumbers. These cultivars are able to set seedless fruit when pollinating insects are absent, but will set fruit with seed when pollinated. The long fruit of these cultivars is often misshapen if pollinated, being enlarged at the blossom end where the seed are located (den Nijs and Miotay, 1991). Homozygous gynoecious cultivars reduce the likelihood of pollination and misshapen fruit, since they produce no pollen.

Homozygous gynoecious hybrid cucumber seed has been produced by crossing two gynoecious lines after one parent has been treated with a growth regulator to induce male flower formation. Hybrid seed produced in this way can be expensive, but the cost may be justified for greenhouse cucumbers since their value is high and the seed requirements are nominal.

Gynoecious × Hermaphroditic Hybrids

Kubicki (1965, 1970) suggested that hybrid cucumber seed homozygous for gynoecious sex expression could be produced by crossing gynoecious and hermaphroditic lines. Hermaphroditic cucumbers, having only perfect flowers, are due to the interaction of genes F and m. Since both parents of the gynoecious × hermaphroditic cross are homozygous for F, the F_1 hybrid will also be homozygous for that gene and therefore will likely be stable for gynoecious sex expression. Mulkey and Pike (1972) found that gynoecious hybrids heterozygous for m, produced by crossing gynoecious × hermaphroditic lines, were more stable for gynoecious sex expression than their gynoecious parents.

The heterozygosity for gene m of gynoecious × hermaphroditic F_1 hybrids should be of no adverse consequence since this gene is recessive. The hybrid will not have the short fruit character that is associated with the m/m homozygote.

Kubicki (1965) bred hermaphroditic cucumbers with the ability to set parthenocarpic fruit, and suggested that they could be used as parents of hybrid cultivars not requiring pollination. Pike and Mulkey (1971) produced completely gynoecious hybrids by crossing gynoecious and hermaphroditic lines. They suggested that the lack of male flowers on gynoecious × hermaphroditic hybrids should make it possible to develop parthenocarpic cutivars producing only seedless fruit.

Staub, Balgooyen, and Tolla (1986) found that gynoecious cucumber hybrids that were genetically similar, except for one of each pair being hetereozygous for m, did not differ in yield or quality. Franken (1981), however, considered it advantageous for gynoecious cucumber hybrids to be heterozygous for m. He reported that hybrids homozygous for F and heterozygous for m were equal to or higher in yield than F/F or $F/+$ hybrids that were homozygous dominant for m.

Kubicki (1965) suggested that 3-way hybrid cultivars could be produced by the cross (gynoecious × hermaphroditic) × monoecious. Mulkey and Pike (1972) concluded that the greater stability of female sex expression of F/F $m/+$ hybrids would be desirable for producing 3-way hybrid seed with a monoecious or hermaphroditic pollinator. The 3-way cross with a hermaphroditic line, however, will likely segregate 1:1 for the undesirably short fruit character that is associated with *m/m*.

Tasdighi and Baker (1981) tested 102 cucumber hybrids of gynoecious, monoecious, hermaphroditic, and androecious lines. The most productive hybrids were crosses of gynoecious × androecious and gynoecious × hermaphroditic lines, followed by the 3-way cross of (gynoecious × hermaphroditic) × androecious.

Kubicki (1965, 1970) advocated the use of complementary hermaphroditic lines to maintain gynoecious inbreds used as parents of hybrid cucumber cultivars. He suggested using the backcross method to incorporate *m* into homozygous gynoecious lines. The resulting hermaphroditic line could be used as the pollen parent in crosses with its nearly isogenic gynoecious counterpart to reproduce a homozygous gynoecious line without the need for growth regulator treatment. Seed produced in this way can be used for the maternal parent of gynoecious hybrid cultivars.

Several years after proposals were made to produce gynoecious × hermaphroditic hybrid cucumber seed, Frankel and Galun (1977) and Staub, Balgooyen, and Tolla (1986) stated that this method had not been used commercially on a large scale. The same still seems true today.

Gynoecious × Androecious Hybrids

Androecious cucumber plants, which have only male flowers due to a single recessive gene (*a*) and modifiers, can be used as the male parent of a hybrid. If the female parent produces no staminate flowers because of its gynoecious sex expression, and the male parent has no fruit because it lacks pistillate flowers, then all the seed produced by open pollination should be hybrid.

Scott and Baker (1976) compared androecious and monoecious pollinators in crosses with gynoecious × hermaphroditic hybrids. The 3-way crosses were consistently more female when the pollinator was androecious. They found that gynoecious × androecious single cross hybrids tended to be more highly female than a commercial gynoecious × monoecious hybrid. They suggested that yield of pickling cucumbers for once-over mechanical harvest could be increased by the use of androecious instead of monoecious pollen parents of gynoecious hybrids. Tasdighi and Baker (1981) also found that hybrids with androecious pollen parents were more highly female and higher yielding than those with monoecious pollen parents.

Niego and Galun (1988) proposed producing hybrid cucumber seed by open pollination of gynoecious and androecious lines. They suggested mixing the seed for the gynoecious and androecious parents and planting it in the same row of the hybrid seed production field. They used 1:4 and 1:10 ratios of androecious to gynoecious seed, and obtained the greatest seed yield with the 1:10 ratio. To increase seed of the androecious parent, they recommended that it be treated with the growth regulator ethephon to induce pistillate flowers.

Several advantages of this method of producing gynoecious hybrid cucumber seed were cited by Niego and Galun (1988)–(1) Bees tend to systematically pollinate plants in the same row rather than travel from one row to another, and will more efficiently cross pollinate the parents when they are grown in the same row; (2) The entire seed production field is planted with the same mixture of seed, and all fruit produced in that field should contain only hybrid seed; (3) A higher proportion of gynoecious plants is possible in the seed production field with an androecious pollinator in the same row than is customary when a monecious pollinator line is grown in a separate row. The increased number of gynoecious plants and better cross pollination could result in increased yield of hybrid seed. Niego, Galun, and Levy (1989) reported 25% increased yield of hybrid seed with this method.

Niego and Galun (1988) discussed one disadvantage of this proposed system, the additional time required to breed androecious sex expression into the pollen parent of a hybrid. Another possible disadvantage, if there is any outcrossing in the previous generation or if an environmental factor causes the androecious parent to produce some pistillate flowers, is that its seed might become mixed with the hybrid seed. When the pollen parent is grown in a separate row, it can be removed from the field before producing seed, but this is less easily accomplished when both parents are in the same row.

The concept of blending seed of a gynoecious and a pollinator line and planting them in the same row of a hybrid seed production field was previously considered by Kubicki (1970). He suggested planting 80% seed of a gynoecious cucumber inbred and 20% seed of a monoecious pollinator. He recommended that the gynoecious and monoecious parental lines could differ for the *B* gene for spine color so that the fruit could be distinguished. Hybrid seed should be harvested from only the gynoecious parent, since the monoecious parent produces nonhybrid seed.

GROWTH REGULATORS
TO MAINTAIN GYNOECIOUS LINES

Since the maternal parent of a gynoecious cucumber hybrid normally produces few or no male flowers, open- or hand-pollination cannot be relied on to increase its seed stocks. The commercial production of gynoecious hybrid cucumber seed was made possible only when it was discovered that gynoecious inbreds could self reproduce if a growth regulator is applied to induce male flower formation.

Peterson and Anhder (1960) and Galun (1973) discovered that gibberellic acid would promote male flower formation on gynoecious cucumber lines. Pike and Peterson (1969) found that a mixture of gibberellins 4 and 7 were more effective than GA$_3$ for maintaining gynoecious lines. Different gynoecious inbreds may vary in response to GA application (Shifriss and George, 1964).

GA was initially used to maintain gynoecious lines used for producing hybrid seed. A problem often arose, however, in that not all of the treated gynoecious plants produced staminate flowers. Genetic variation in gynoecious lines results in plants with the strongest tendency towards maleness being more likely to respond to GA treatment by producing male flowers than plants with a higher degree of female sex expression. Consequently, there could be a genetic shift in a gynoecious line after several generations of using GA to increase seed, and hybrids produced with an altered gynoecious line may produce an increased number of monoecious plants. Also, the number of male flowers induced by GA may not be sufficient for using a gynoecious line as a male parent of a homozygous gynoecious hybrid (Tolla and Peterson, 1979).

Because of these problems, many seedsmen now use a silver compound rather than gibberellin to maintain gynoecious lines. Silver ions inhibit ethylene action and therefore promote male sex expression of gynoecious cucumbers (Beyer, 1976). Tolla and Peterson (1979) compared silver nitrate and gibberellins 4 and 7 for inducing staminate flowers on a gynoecious inbred. They concluded that silver nitrate induced significantly more staminate flowers.

Silver thiosulfate is now used by many seedsmen to increase seed of gynoecious cucumber lines. It effectively induces male flowering of gynoecious plants for an extended period, is less phytotoxic than silver nitrate, and does not cause the excessive stem elongation or malformed male flowers that is characteristic of gibberellin application (den Nijs and Visser, 1980).

Gibberellins and several other growth regulating chemicals did not alter the sex expression of *Cucurbita foetidissima* (Scheerens et al., 1988). Application of aminoethoxyvinylglycine, however, induced male flower formation on gynoecious *C. foetidissima* plants, and it was suggested that AVG could be used to maintain gynoecious inbreds of this species.

GROWTH REGULATORS
TO FACILITATE HYBRID SEED PRODUCTION

A number of growth regulating chemicals are known to promote female sex expression of cucurbits. The most practical growth regulator to use in the production of hybrid cucurbit seed is ethephon, since it can have such a persistent effect on female flower formation of some species. When properly applied to the female parent of a hybrid cultivar of those species, it can cause the plants to produce only female flowers for such an extended period that hybrid seed can be produced by open pollination.

Robinson, Shannon, and de la Guardia (1969) reported that monoecious cucumbers treated with ethephon in the seedling stage produced only female flowers at the first 18 nodes, but untreated plants developed several male flowers at each node. Treated plants had good fruit set and seed production, and it was suggested that hybrid cucumber seed production could be accomplished by ethephon treatment of the naturally pollinated maternal parent.

Rudich, Kedar, and Halevy (1970) found that the combined application of ethephon and Alar inhibited male flower formation of melon. Treated andromonoecious plants still produced perfect flowers, however, and therefore could not be used to produce 100% hybrid seed without emasculation. Treated monoecious melon plants produced only female flowers for up to three weeks. They concluded that this treatment of monoecious plants might be feasible for producing hybrid melon seed, but the association of monoecy with fruit shape could be a problem since the hybrid would likely have elongated fruit.

Lee and Janick (1978) reported that ethephon application to andromonoecious melons promoted female sex expression and increased hybrid seed production by natural field crossing. Treated plants were hermaphroditic, with varying degrees of anther abortion, and considerable selfing and sibbing occurred. The amount of hybrid seed was increased by ethephon treatment from 12.0 to 27.4% in their tests, and from 16.4 to as much as 56.3% in tests by Alvarez (1988). This is a significant increase in the proportion of hybrid seed, but ethephon treatment of andromonoecious melons is not adequate for the commercial production seed of entirely F_1 hybrid seed without emasculation.

Cucurbita pepo, *C. maxima*, and *C. moschata* plants produced only female flowers for an extended period after being treated with ethephon in the seedling stage (Robinson, Whitaker, and Bohn, 1970). Shannon and Robinson (1979) reported that two applications of 400 to 600 ppm ethephon was satisfactory for the commercial production of hybrid summer squash seed. Treated plants produced normal amounts and quality of hybrid seed.

Boggini et al. (1972) and Borghi and Pironi (1974) had similar results. They suggested applying ethephon at the rate of 400 ppm active ingredient to the maternal *C. pepo* parent in the 4-5 leaf stage and again seven to ten days later. They recommended growing one row of the pollinator line for every ten rows of the seed parent treated with ethephon when producing F_1 seed of *Cucurbita pepo*. Lercari and Tesi (1976) found no difference in seed yield or germination when there were from one to seven rows of *C. pepo* plants treated with ethephon for each row of the pollinator line.

Hume and Lovell (1981) compared ethephon treatment with hand pollination and defloration for the production of hybrid seed of *Cucurbita maxima*. Ethephon treatment reduced the amount of labor required by 90%, and resulted in greater production of high quality hybrid seed. Korzeniewska and Niemirowicz-Szczytt (1996) reported that hybrid seed yield was generally higher for *Cucurbita maxima* treated with ethephon and pollinated by bees than for manual pollinations, and there was more than 80% savings in labor with the ethephon method. Boggini et al. (1972) reported 20% increase in hybrid seed production with ethephon treatment.

Different cucurbit species and cultivars vary in their response to ethephon. In general, female sex expression is promoted more for monoecious squash and cucumber than for melon or watermelon plants. Bush cultivars of squash are more responsive than vining types, and cultivars with an inherently high degree of female sex expression are easier to convert to gynoecy than highly male types.

The principle use of ethephon today in cucurbit seed production is for F_1 hybrids of summer squash. Wittwer estimated in 1974 that approximately 250,000 pounds of hybrid squash seed was produced with ethephon treatment. Ethephon is also used commercially to produce seed of monoecious cucumber hybrids.

MALE STERILITY

Single recessive genes for male sterility are known for cucumber, melon, squash, and watermelon (Robinson et al., 1976). Since no cytoplasmic factor is known in cucurbits to interact with any of the male sterile genes, male sterile lines are maintained by crossing plants heterozygous with those homozygous for the male sterile gene, using the latter as the female parent. The progeny segregates 1:1 for male sterility and must be rogued to remove the fertile plants when used to produce hybrid seed. In watermelon, the *gms* gene for male sterility is associated with glabrous foliage, and male sterile plants can be identified in the seedling stage.

Foster (1970) proposed that hybrid melon seed could be produced by open pollination of monoecious male sterile plants. He determined that male fertile monoecious plants produced 58% hybrid seed when grown with normal melon plants. He calculated that if the monoecious maternal parent segregated 1:1 for male sterility, then the population should produce a proportion of hybrid seed intermediate to that amount and the nearly 100% hybrid seed that male sterile plants produce. This assumption was correct, for he determined that a monoecious line segregating 1:1 for male sterility produced 78% hybrid seed.

The use of male sterility in hybrid seed production for cucurbits has been limited because of the labor required to rogue segregating male sterile lines, the possibility of some selfing of the maternal parent of a hybrid if some male fertile plants are not rouged before they flower, and the reduced seed yield when seed is harvested from only 50% of the plants. Many male sterile mutants cannot be identified until flower buds are visible. Rogueing in this late stage of development may result in some male fertile plants flowering and providing pollen to produce nonhybrid seed before they are removed. In addition, detrimental effects are associated with some male sterile genes. Male sterile cucumber plants can have reduced seed production (Shifriss, 1945). Zhang, Gabert, and Baggett (1994) concluded that the use of a pollen sterile cucumber mutant in F_1 hybrid seed production is not practical. Bhattacharya, Kato, and Jodo (1970) observed that male sterile melons had smaller fruit than normal in the winter greenhouse. More and Seshadri (1998) concluded that the practicality of large scale seed production of hybrid melons by using male sterility is not yet worked out.

In the U.S., the principle use of male sterility in hybrid cucurbit seed production has been for winter squash, *Cucurbita maxima*. Male sterility has been used in China for producing hybrid watermelon seed (Zhang and Rhodes, 1993).

TRIPLOID HYBRIDS

Kihara (1951) developed a method of producing hybrid watermelons that are nearly seedless. Tetraploid lines are obtained by colchicine treatment, and are then maintained by self or sib pollination. To produce hybrid seed, the tetraploid line is pollinated by hand with diploid pollen. The triploid progeny is highly sterile and produces few if any seed. Three-way hybrid cultivars of seedless watermelon have been developed by crossing a tetraploid line with a diploid F_1 hybrid (Kanda, 1984a).

Triploid hybrid watermelon seed produced by hand pollination is expensive because of the cost of labor and the low seed yield of the tetraploid parent. Wall (1960) stated that $4n$ watermelon plants produce only 25 to 40% as much seed as comparable diploids. Mohr (1986) concluded that the cost of triploid hybrid seed was approximately 20 times that of open pollinated diploid seed.

To avoid the high cost of hand pollinated seed for seedless watermelons, Wall (1960) proposed that open pollination be used. He determined that the open pollinated progeny of tetraploid plants paired with diploids produced 83.6% triploid seed. He attributed this unusually high rate of natural cross pollination for watermelon to the reduced self fertility of the tetraploid maternal parent. Shimotsuma (1962) found that the amount of $3n$ hybrid seed produced by open pollination of $4n$ plants grown with $2n$ plants was highest with a ratio of 3 tetraploid:1 diploid plant in the seed production field. The triploid seed produced by open pollinated tetraploid plants can be separated from tetraploid seed on the basis of seed thickness and weight (Shimotsuma, 1959).

Seed of a diploid pollinator blended with that of a triploid cultivar may improve pollination and fruit set of seedless watermelon. Kihara (1951) recommended a ratio of 4 to 5 triploid to 1 diploid seed. It is best to select a pollinator with fruit easily distinguished from the hybrid, such as differing for fruit striping, so that the fruit can be easily separated before marketing.

Triploid hybrids of *Cucumis melo* that are nearly seedless have been produced (Kihara, 1958), but they have had no commercial importance. Seedless fruit are less advantageous for melon than for watermelon, since the seed are not embedded in the flesh and are easily separated.

INTERSPECIFIC HYBRID CULTIVARS

The interspecific hybrid of *Cucurbita maxima* and *C. moschata* is possible, but is often not easily obtained. Fruit set is generally quite low for many crosses of these two species, the occasional fruit produced may have few seed or none, and the hybrid is usually highly sterile (Whitaker and Davis, 1962). Thus, it is rather remarkable that Japanese seedsmen have produced seed of this interspecific hybrid so efficiently that it is marketed commercially.

The first winter squash cultivar of the *C. maxima* × *C. moschata* interspecific cross was 'Tetsukabuto'. It is a cross of 'Delicious' (*C. maxima*) and 'Kurokawa No. 2', an early cultivar of *C. moschata* (Kanda, 1984b). The success in producing seed economically for 'Tetsakuboto' and other *C. maxima* × *C. moschata* cultivars is apparently due to good combining ability, the result of selecting parents that produce more hybrid seed than many other parental combinations of these two species.

Particular care should be taken to guard against pollen of the maternal parent when producing seed of this interspecific hybrid. Pollen tubes of the maternal parent grow more quickly in the style of that parent than the pollen tubes of the other species (Kanda, 1984b), and gamete competition will result in a disproportional amount of contamination of the hybrid seed with seed of the maternal parent if there is any selfing or sibbing. Kanda (1984b) recommended applying NAA to the pistillate flower when making the interspecific cross to improve fruit set.

Both *C. maxima* and *C. moschata* are monoecious, producing many more male than female flowers on the same plant, but the interspecific hybrid is gynoecious or predominantly female. An open pollinated cultivar is often planted with *C. maxima* × *C. moschata* hybrids in order to provide pollen.

C. maxima × *C. moschata* hybrid cultivars are generally quite productive, presumably due to heterosis and the abundance of pistillate flowers. Their vigor is so exceptional that they are utilized as root stocks for grafting cucumber, watermelon, and melon plants (Kanda, 1984b).

Interspecific hybrids have been used as rootstocks in the production of hybrid seed. The exceptional vigor of the interspecific hybrid of *Cucurbita maxima* and *C. moschata* has been utilized in Japan to increase cucumber seed production. The *Cucurbita* interspecific hybrid rootstock increased cucumber seed production by 50 to 60% (Takahashi, 1987).

USE OF MARKER GENES IN HYBRID SEED PRODUCTION

Marker genes can be used to identify hybrid plants produced by open pollination. If the maternal parent is homozygous for a recessive marker gene, its open pollinated progeny will be mutant if selfed or sibbed but it will be heterozygous, and therefore normal in appearance, if hybrid. This permits the grower to discard the mutants in mixed populations, leaving only hybrid plants in the field.

A marker gene can also be used when increasing seedstocks of the inbred parents of a hybrid cultivar. Kubicki (1970) proposed that nearly isogenic gynoecious and hermaphroditic lines could be planted together to increase seed of each line by open pollination. The short fruit character associated with the m/m genotype of the hermaphroditic line can be used a marker to distinguish fruit of the isogenic lines. At harvest, the long fruit of the gynoecious line can be separated from the short fruit of the hermaphroditic line, and its seed will produce gynoecious ($F/F\ m/+$) plants that can be used as the maternal parent of a hybrid. Open pollinated seed from the short fruit of the hermaphroditic plants ($F/F\ m/m$) will be selfed or sibbed, since the other line produces no male flowers, and can be used to maintain that hermaphroditic line.

A recessive gene (*nl*) for nonlobed leaves has been used in China (Zhang and Rhodes, 1993) to identify true hybrids of watermelon in mixed populations. Mohr, Blackhurst, and Jensen (1955) obtained 20.6 to 36.0% hybrid seed when open pollinated seed was harvested from this mutant when grown at different plant arrangements with normal plants.

Another recessive gene of watermelon, *gms*, can be used to facilitate the production of hybrid seed by open pollination (Watts, 1962). Mutant plants have a seedling marker, glabrous foliage, that is associated with male sterility. Reduced female fertility and instability of male sterility of the *gms* mutant, however, has deterred its use in the commercial production of hybrid watermelon seed (Zhang et al., 1996). A recessive seedling marker gene is linked to a male sterile gene of cucumber (Whelan, 1974), but this association has not been used in the commercial production of hybrid seed.

There are no known undesirable effects associated with the *ms* male sterile gene of watermelon, and it has been used to produce F_1 hybrid seed in China. Zhang et al. (1996) proposed that a non-linked virescent gene, *dg*, could be combined with *ms* to facilitate hybrid seed production by open pollination. The seedling marker gene *dg* would make possible early tests of hybrid seed purity, since hybrids would have normal phenotype and selfs or sibs would be chlorotic in the seedling stage. Also, *dg* would indicate if the male sterile line was correctly maintained or if there were contamination by outcrossing or mechanical mixture.

Wall (1960) produced a high proportion of triploid hybrid watermelon seed by harvesting open pollinated seed from tetraploid plants paired with diploids. The parents differed for exocarp genes that enable a grower to distinguish the hybrid seedless fruit from nonhybrid fruit.

Kubicki (1970) suggested that the *B* gene of cucumber could be used as a marker to distinguish open pollinated fruit with hybrid seed, produced by the gynoecious parent, from the fruit of the monoecious pollinator which has nonhybrid seed. This would mean, however, that the hybrid would have black fruit spines, and this is undesirable for slicing cucumber cultivars.

Foster (1970) used a recessive seedling marker gene, *gl* (glabrous), to identify melon hybrids produced by open pollination. He determined that using a monoecious instead of an andromonoecious line as the maternal glabrous parent more than doubled the proportion of hybrid seed produced, 58.4 vs. 24.6%. The percentage hybrid seed produced was greater early in the season than later, and was significantly increased if the maternal parent segregated for male sterility. He used two seedling marker genes of melon to demonstrate (1968b) that mixed populations containing as little as 25% hybrid seed could be rogued to produce a perfect stand of hybrid plants with little increase in cost.

A mixed population with a relatively high frequency of nonhybrid plants may be satisfactory in some situations. But when uniformity is essential, such as for once-over mechanical harvest of cucumber, such heterogenous populations are unacceptable. Also, many growers are unwilling to take the time and loss of stand to remove nonhybrid plants from mixed plantings. The use of marker genes to permit hybrid seed production by open pollination has not become commercially important.

REFERENCES

Allard, R. W. (1960). *Principles of Plant Breeding.* John Wiley, NY. 485 pp.

Alvarez, H. (1988). Muskmelon hybrid seed production through ethephon-induced feminization in andromonoecious cultivars. *Proc. Eucarpia Mtg. Cucurbit Breed. Genet.*, Paris. pp. 89-97.

Amaya, A. T. and S. G. Ortega. (1996). High yields of summer squash lines and hybrid combinations. *Cucurbit Genet. Coop. Rept.* 19: 78-80.

Ambrose, J. T., J. R. Schultheis, S. B. Bambara, and W. Mangum. (1995). An evaluation of selected commercial bee attractants in the pollination of cucumbers and watermelons. *Amer. Bee J.* 135: 267-271.

Bailey, L. H. (1890). Experiments in crossing cucurbits. *Cornell Agr. Expt. Sta. Bul.* 25.

Barnes, W. C. (1947). Cucumber breeding methods. *Proc. Amer. Soc. Hort. Sci.* 49: 227-230.

Barnes, W. C. (1966). Development of multiple disease resistant hybrid cucumbers. *Proc. Amer. Soc. Hort. Sci.* 89: 390-393.

Bemis, W. P., J. W. Berry, C. Weber, and T. W. Whitaker. (1978). The buffalo gourd: a new potential horticultural crop. *HortScience* 13: 235-240.

Beyer, E. Jr. (1976). Silver ion: a potent antiethylene agent in cucumber and tomato. *HortScience* 11: 195-196.

Bhattacharya, A., M. Kato, and S. Jodo. (1970). Use of male sterility for heterotic effect in F_1 hybrids of muskmelon (*Cucumis melo* L.). *Mem. College Agr. Ehime Univ.* 15: 21-30.

Boggini, G., B. Borghi, T. Maggiore, and C. P. Treccani. (1972). Produzione di seme ibrido nello zucchino (*Cucurbita pepo*). *Riv. Agron.* 6: 165-170.

Bohn, G. W. and G. N. Davis. (1957). Earliness in F_1 hybrid muskmelons and their parent varieties. *Hilgardia* 26: 453-471.

Borchers, E. A. and R. T. Taylor. (1988). Inheritance of fruit bitterness in a cross of *Cucurbita mixta* × *C. pepo. HortScience* 23: 603-604.

Borghi, B., T. Maggiore, G. Boggini, and F. Bonali. (1973). Inbreeding depression and heterosis in *Cucurbita pepo* evaluated by means of diallelical analysis. *Genet. Agraria* 27: 415-431.

Borghi, B. and W. Pironi. (1974). Evaluation of heterosis in *Cucurbita pepo* L. In: *Heterosis in Plant Breeding*, eds. A. Jánossy and F. G. H. Lupton. Proc. 7th Eucarpia Congr. Elsevier Sci. Publ. Co., Amsterdam. pp. 219-226.

Bowers, J. L. (1967). Hand pollination of cantaloupes. *Veg. Impr. Newsl.* 9: 6.

Brewer, J. W. (1974). Pollination requirements for watermelon seed production. *J. Apic. Res.* 13: 207-212.

Buchinger, A. (1948). Kürbiszüchtung. *Bodenkultur.* 2: 10-27.

Burrell, P. C. and T. W. Whitaker. (1939). The effect of indol-acetic acid on fruit-setting in muskmelons. *Proc. Amer. Soc. Hort. Sci.* 37: 829-830.

Bushnell, J. W. (1920). The fertility and fruiting habit in Cucurbita. *Proc. Amer. Soc. Hort. Sci.* 17: 47-51.

Bushnell, J. W. (1922). Isolation of uniform tupes of "Hubbard" squash by inbreeding. *Proc. Amer. Soc. J. Hort. Sci.* 19: 139-144.

Chadha, M. L. and K. S. Nandpuri. (1980). Hybrid vigour studies in muskmelon. *Indian J. Hort.* 37: 276-282.

Chekalina, I. N. (1975). Effect of inbreeding on variability of winter squash. *Genetika* 12 (5): 45-49.

Choudhury, B. and B. Singh. (1971). Two high yielding bottlegourd hybrids. *Indian Hort.* 16: 15, 32.

Choudhury, B. and M. R. Thakur. (1965). Inheritance of sex forms in *Luffa. Indian J. Genet. Plant Breed.* 25: 188-197.

Collison, C. H. and E. C. Martin. (1970). Competitive plants that may affect the pollination of pickling cucumbers by bees. *Amer. Bee J.* 110: 262.

Collison, C. H. and E. C. Martin. (1973). The effects of overhead irrigation on the pollination of pickling cucumbers. *Pickle Pak Science* 3: 1-3.

Collison, C. H. and E. C. Martin. (1974). Pickle research at Michigan State University–1973-1974. *Michigan State Univ. Res. Rpt.* 277: 3-6.

Cummings, M. B. (1904). Fertilization problems: a study of reciprocal crosses. *Maine Agr. Expt. Sta. Bul.* 104: 81-100.

Cummings, M. B. and E. W. Jenkins. (1928). Pure line studies with ten generations of Hubbard squash. *Vermont Agr. Expt. Sta. Bul.* 28. 29 pp.

Cummings, M. B. and W. C. Stone. (1921). Yield and quality in Hubbard squash. *Vermont Agr. Expt. Sta. Bull.* 222. 48 pp.

Curtis, L. C. (1939). Heterosis in summer squash (*Cucurbita pepo*) and the possibilities of producing F_1 hybrid seed for commercial planting. *Proc. Amer. Soc. Hort. Sci.* 37: 827-828.

Curtis, L. C. (1941). Comparative earliness and productiveness of first and second generation summer squash (*Cucurbita pepo*) and the possibilities of using the second generation for commercial planting. *Proc. Amer. Soc. Hort. Sci.* 38: 596-598.

Devadas. V. S. and S. Ramadas. (1995). Combining ability for seed yield and quality parameters in bitter gourd (*Momordica charantia* L.). *Indian J. Genet. Plant Breed.* 55: 41-45.

Dhesi, N. S., K. S. Nandpuri, and J. Chand. (1964). Hybrid vigour in Indian squash. *Indian J. Genet. Pl. Breed.* 24: 57-59.

Dixit, J. and Kalloo. (1983). Heterosis in muskmelon (*Cucumis melo* L.). *Haryana Agr. Univ. J . Res.* 13: 549-553.

Doijode, S. D. and U. V. Sulladmath. (1983). Preliminary studies on heterosis in pumpkin (*Cucurbita moschata* Poir.). *Mysore J. Agr. Sci.* 13: 30-34.

Edgecomb, S. W. (1946). Honeybees as pollinators in the production of hybrid cucumber seed. *Amer. Bee J.* 86: 147.

El-Shawaf, I. I. S. and L. R. Baker. (1981). Inheritance of parthenocarpic yield in gynoecious pickling cucumber for once-over mechnical harvest by diallel analysis of six gynoecious lines. *J. Amer. Soc. Hort. Sci.* 106: 359-364.

Erwin, A. T. and E. S. Haber. (1929). Species and varietal crosses in cucurbits. *Iowa Agr. Expt. Sta. Bull.* 263: 343-372.

Foster, R. E. (1967). F_1 hybrid muskmelons, I. Superior performance of selected hybrids. *Proc. Amer. Soc. Hort. Sci.* 91: 390-395.

Foster, R. E. (1968a). F_1 hybrid muskmelons, III. Field production of hybrid seed. *Proc. Amer. Soc. Hort. Sci.* 92: 461-464.

Foster, R. E. (1968b). F_1 hybrid muskmelons, IV. Rogueing-thinning to pure stands from mixed seed. *J. Ariz. Acad. Sci.* 5: 77-79.

Foster, R. E. (1970). F_1 hybrid muskmelons, V. Monoecism and male sterility in commercial seed production. *J. Hered.* 59: 205-207.

Foster, R. E. and M. D. Levin. (1967). F_1 hybrid muskmelons, II. Bee activity in seed fields. *J. Ariz. Acad. Sci.* 4: 222-225.

Frankel, R. and E. Galun. (1977). *Pollination Mechanisms, Reproduction and Plant Breeding.* Springer-Verlag, Berlin. 281 pp.

Franken, S. (1981). Genetic investigations of determinate pickling cucumber (*Cucumis sativus* L.). 2. Hermaphroditism and its use in hybrid breeding. *Z. Pflanzenzüchtg.* 186:136-147.

Galun, E. (1973). The use of genetic sex types for hybrid seed production in Cucumis. In: *Agricultural Genetics*, ed. R. Moav. John Wiley, NY. pp. 23-56.

Ganesan, J. (1976). Natural cross pollination in musk melon (*Cucumis melo* L.). *Annalmalai Univ. Agr. Res. Ann.* 6: 49-52.

George, R. A. T. (1985). *Vegetable Seed Production.* Longman, London.

Ghaderi, A. and R. L. Lower. (1978). Heterosis and phenotypic stability of F_1 hybrids in cucumber under controlled environment. *J. Amer. Soc. Hort. Sci.* 103: 275-278.

Ghaderi, A. and R. L. Lower. (1979a). Analysis of generation means for yield in six crosses of cucumber. *J. Amer. Soc. Hort. Sci.* 104: 567-572.

Ghaderi, A. and R. L. Lower. (1979b). Heterosis and inbreeding depression for yield in populations derived from six crosses of cucumber. *J. Amer. Soc. Hort. Sci.* 104: 564-567.

Grebenscikov, I. (1967). Zur quantitativ-genetischen Analyse der Ertragskomponenten beim Kürbis. Teil 3. Reziproke Kreuzung zweier stark verschiedener Typen von Cucurbita maxima. *Kulturpflanze* 15: 57-74.

Grebenscikov, I. (1975). Notulae Cucurbitologicae VIII. Zur Frage der Reziprokenunterschiede bei den quantitativen Ertragsmerkmalen vom Kürbis. *Kulturpflanze* 23: 139-155.

Groff, D. (1993). Monoecious hybrid blenders are improvement for pickle industry. *Asgrow Seed Co. UJ* 7671, 2 pp.

Gubin, A. F. (1945). Bee training for pollination of cucumbers. *Bee World* 26: 34-34.

Haber, E. S. (1928). Inbreeding the Table Queen (Des Moines) squash. *Proc. Amer. Soc. Hort. Sci.* 25: 111-114.

Hawthorn, L. R. and L. H. Pollard. (1954). *Vegetable and Flower Seed Production.* Blakiston Co., NY. 626 pp.

Hayase, H. and T. Ueda. (1956). *Cucurbita* cross. IX. Hybrid vigour of reciprocal F_1 crosses in *Cucurbita maxima. Hokkaido Natl. Agr. Expt. Sta. Res. Bull.* 71: 119-128.

Hayes, H. K. and D. F. Jones. (1916). First generation crosses in cucumbers. *Conn. Agr. Expt. Sta. Ann. Rept.* 1916: 319-322.

Hollar, L. A. 1998. Personal communication.

Hoshino, H. and M. Furuich. (1984). Cucumber. In: *Vegetable Seed Production Technology of Japan Elucidated with Respective Variety Development Histories, Particulars*, ed. S. Shinohara. Shinohara's Authorized Agr. Consult. Engineer Office, Nishiooi, Japan. Vol 1, pp. 340-360.

Hume, R. J. and P. H. Lovell. (1981). Reduction of the cost involved in hybrid seed production of pumpkins (*Cucurbita maxima* Duchesne). *New Zealand J. Exptl. Agr.* 9: 209-210.

Hutchins, A. E. (1938). Some examples of heterosis in the cucumber, *Cucumis sativus* L. *Proc. Amer. Soc. Hort. Sci.* 36: 660-664.

Hutchins, A. E. and F. E. Croston. (1941). Productivity of F_1 hybrids in the squash, *Cucurbita maxima. Proc. Amer. Soc. Hort. Sci.* 39: 332-336.

Ivanoff, S. S. (1947). Natural self-pollination in cantaloupes. *Proc. Amer. Soc. Hort. Sci.* 50: 314-316.

James, E., J. H. Massey, and W. L. Corley. (1960). Effect of a plant arrangement on cross-pollination of muskmelons. *Proc. Amer. Soc. Hort. Sci.* 75: 480-484.

Jenkins, J. M. Jr. (1942). Natural self-pollination in cucumbers. *Proc. Amer. Soc. Hort. Sci.* 40: 411-412.

Kadam, P. Y., U. T. Desai, and P. N. Kale. (1995). Heterosis studies in ridge gourd. *J. Maharashtra Agr. Univ.* 20: 119-120.

Kanda, T. (1984a). Watermelon: F_1 hybrid seed production. In: *Vegetable Seed Production Technology of Japan Elucidated with Respective Variety Development Histories, Particulars*, ed. S. Shinohara. Shinohara's Authorized Agr. Consult. Engineer Office. Nishiooi, Japan. Vol I, pp. 318-339.

Kanda, T. (1984b). Squash. In: *Vegetable Seed Production Technology of Japan Elucidated with Respective Variety Development Histories, Particulars*, ed. S. Shinohara. Shinohara's Authorized Agr. Consult. Engineer Office. Nishiooi, Japan. Vol I, pp. 395-426.

Kasrawi, M. A. (1994). Heterosis and reciprocal differences for quantitative traits in summer squash (*Cucurbita pepo* L.). *J. Genet. Breed.* 48: 399-403.

Kihara, H. (1951). Triploid watermelons. *Proc. Amer. Soc. Hort. Sci.* 58: 217-230.

Kihara, H. (1958). Breeding of seedless fruits. *Seiken Ziho* 9: 1-6.

Kim, M. S., Y. K. Kim, and H. D. Chung. (1996). Combining ability of fruit yield and quantitative characters in muskmelon (*Cucumis melo* L.). *J. Korean Soc. Hort. Sci.* 37: 657-661.

Kitroongruang, N., W. Poo-swang, and S. Tokumasu. (1992). Evaluation of combining ability, heterosis and genetic variance for plant growth and fruit quality characteristics in Thai-melon (*Cucumis melo* var. *acidulus* Naud.). *Sci. Hort.* 50: 79-87.

Kooistra, E. (1967). Femaleness in breeding glasshouse cucumbers. *Inst. Hort. Plant Breed. Med.* 269. 17 pp.

Korzeniewska, A. and K. Niemirowicz-Szczytt. (1993). Combining ability and heterosis effect in winter squash (*Cucurbita maxima* Duch.). *Genet. Polonica* 34: 259-272.

Korzeniewska, A. and K. Niemirowicz-Szczytt. (1996). Evaluation of the method for seed production by three F₁ hybrids of winter squash (*Cucurbita maxima* Duch.) using ethephon. *Proc. VI Eucarpia Mtg. Cucurbit Genet.*, Malaga, Spain. pp. 82-87.

Kubicki, B. (1962). Inheritance of some characters in muscmelons (*Cucumis melo* L.). *Genet. Polonica* 3: 265-274.

Kubicki, B. (1965). New possibilities of applying different sex types in cucumber breeding. *Genet. Polonica* 6: 241-250.

Kubicki, B. (1966). Genetic basis for obtaining gynoecious muskmelon lines and the possibility of their use for hybrid seed production. *Genet. Polonica* 7: 27-30.

Kubicki, B. (1970). Cucumber hybrid seed production based on gynoecious lines multiplied with the aid of complementary hermaphroditic lines. *Genet. Polonica* 11: 181-186.

Lana, E. P. (1950). Reciprocal crosses in the squash, *Cucurbita Maxima* Duch. *Minn. Agr. Expt. Sta. Tech. Bul.* 189. 28 pp.

Lawande, K. E. and A. V. Patil. (1989). Heterosis and combining ability studies in bitter gourd. *J. Maharashtra Agr. Univ.* 14: 292-295.

Lawande, K. E. and A. V. Patil. (1990). Studies on combining ability and gene action in bitter gourd. *J. Maharashtra Agr. Univ.* 15: 24-28.

Lee, C. W. and J. Janick. (1978). Muskmelon hybrid seed production facilitated by ethephon. *HortScience* 13:195-196.

Lercari, B. and R. Tesi. (1976). The extent of cross-pollination in *Cucurbita pepo* with flowering habit modified by Ethrel. *Riv. Ortoflorofrutt.* 60: 199-203.

Lippert, L. F. and M. O. Hall. (1972). Hybrid vigor in muskmelon crosses. *Calif. Agr.* 26 (2): 12-14.

Lower, R. L. and M. D. Edwards. (1986). Cucumber breeding. In: *Breeding Vegetable Crops*, ed. M. J. Bassett. Avi Publ. Co. Westport, Connecticut. pp. 173-207.

Lower, R. L., J. Nienhuis, and C. H. Miller. (1982). Gene action and heterosis for yield and vegetative characteristics in a cross between a gynoecious pickling cucumber inbred and a *Cucumis sativus* var. *hardwickii* line. *J. Amer. Soc. Hort. Sci.* 107: 75-78.

Lozanov, P. (1969). Heterosis in certain inter-varietal hybrids of vegetable marrow. *Genetics Plant Breeding* 2: 277-286.

Lumsden, D. (1914.) Mendelism in melons. *New Hampshire Agr. Expt. Sta. Bull.* 172, 58 pp.

Mann, L. K. (1953). Honey bee activity in relation to pollination and fruit set in the cantaloupe *(Cucumis melo)*. *Amer. J. Bot.* 40: 545-553.

Mann, L. K. and J. Robinson. (1950). Fertilization, seed development, and fruit growth as related to fruit set in the cantaloupe *(Cucumis melo L.)*. *Amer. J. Bot.* 37: 685-697.

Martin, E. C. (1970). The use of honey bees in the production of hybrid cucumbers for mechanical harvest. *Arkansas Agr. Expt. Sta. Misc. Publ.* 127:106-109.

McCreight, J. D., H. Nerson, and R. Grumet. (1993). Melon *Cucumis melo* L. In: *Genetic Improvement of Vegetable Crops*, eds. G. Kalloo and B. O. Bergh. Pergamon Press, Oxford. pp. 267-294.

McDonald, M. B. and L. O. Copeland. (1997). *Seed Production Principles and Practices*. Chapman & Hall, NY.

McGregor, S. E. (1976). *Insect Pollination of Cultivated Plants*. U.S. Dept. Agr. Handbook 496. 411 pp.

McGregor, S. E., M. D. Levin, and R. E. Foster. (1965). Honey bee visitors and fruit set of cantaloupes. *J. Econ. Ent.* 58: 968-970.

McGregor, S. E. and F. E. Todd. (1952). Cantaloup production with honey bees. *J. Econ. Ent.* 45: 43-47.

Michelbacher, A. E., R. F. Smith, and P. D. Hurd Jr. (1964). Bees are essential . . . pollination of squashes, gourds and pumpkin. *Calif. Agr.* 18 (5): 2-4.

Miller, C. H. (1976). Effects of blending gynoecious and monoecious cucumber seeds on yield patterns. *HortScience* 11: 428-430.

Mishra, H. N., R. S. Mishra, S. N. Mishra, and G. Parhi. (1994). Heterosis and combining ability in bittergourd *(Momordia charantia)*. *Indian J. Agr. Sci.* 64: 310-313.

Mishra, J. P. and V. S. Seshadri. (1986). Male sterility in muskmelon *(Cucumis melo L.)*: II. Studies on heterosis. *Genet. Agraria* 39: 367-376.

Mohr, H. C. (1986). Watermelon breeding. In: *Breeding Vegetable Crops*, ed. M. J. Bassett. Avi Publ. Col, Westport, CT. pp. 37-66.

Mohr, H. C., H. T. Blackhurst, and E. R. Jensen. (1955). F_1 hybrid watermelons from open-pollinated seed by use of a genetic marker. *Proc. Amer. Soc. Hort. Sci.* 65: 399-404.

More, T. A., V. S. Seshadri, and J. C. Sharma. (1980). Monoecious sex expression in muskmelon, *Cucumis melo* L. *Cucurbit Genet. Coop. Rept.* 3: 32-32.

More, T. A. and V. S. Seshadri. (1980). Studies on heterosis in muskmelon *(Cucumis melo L.)*. *Veg. Sci.* 7: 27-40.

More, T. A. and V. S. Seshadri. (1988). Development of tropical gynoecious lines in cucumber. *Cucurbit Genet. Coop. Rept.* 11: 17.

More, T. A. and V. S. Seshadri. (1998). Improvement and cultivation: muskmelon, cucumber and watermelon. In: *Cucurbits*, eds. N. M. Nayar and T. A. More. Science Publ., Enfield, New Hampshire. pp. 168-186.

More, T. A., V. S. Seshadri, and M. B. Magdum. (1987). Development of gynoecious lines in muskmelon. *Cucurbit Genet. Coop Rpt.* 10: 49-50.

Mulkey, W. A. and L. M. Pike. (1972). Stability of gynoecism in cucumber *(Cucumis sativus* L.) as affected by hybridization with the hermaphrodite 'TAMU 950'. *HortScience* 7: 284-285.

Munger, H. M. (1942). The possible utilization of first generation musk-melon hybrids and an improved method of hybridization. *Proc. Amer. Soc. Hort. Sci.* 40: 405-410.

Munger, H. M. (*1978*). Use of bacterial wilt resistance in hybrid cucumbers. *Veg. Impr. Newsl.* 20: 9.

Munger, H. M. (1985). Near-isogenic lines of several cucumber varieties. *Cucurbit Genet. Coop. Rept.* 8: 4-6.

Munger, H. M. (1993). Breeding for viral disease resistance in cucurbits. In: *Resistance to Viral Diseases of Vegetables*, ed. M. M. Kyle. Timber Press. Portland, Oregon. pp. 44-60.

Munger, H. M., M. M. Kyle, and R. W. Robinson. (1993). Cucurbits. In: *Traditional Crop Breeding Practices: An Historical Review to Serve as a Baseline for Assessing the Role of Modern Biotechnology*. Org. Economic Coop. Development, Paris. pp. 47-60.

Munson, W. M. (1892). Preliminary notes on the secondary effects of pollination. *Maine Agr. Expt. Sta. Ann. Rpt.* 58 pp.

Nath, P. and O. P. Dutta. (1971). Hybridization among muskmelon, snapmelon and longmelon. *Indian J. Hort.* 28: 123-129.

Natti, T. A. and J. B. Loy. (1978). Role of wound ethylene in fruit set of hand-pollinated muskmelons. *J. Amer. Soc. Hort. Sci.* 103: 834-836.

Niego, S. and E. Galun. (1988). A novel procedure for the production of F_1 hybrid cucumber seeds. *Proc. Eucarpia Mtg. Cucurbit Breed. Genet.*, Paris. pp. 99-107.

Niego, S., E. Galun, and M. Levy. (1989). Production of hybrid cucumber seeds. U.S. Patent No. 4,822,949.

Nijs, A. P. M. den and P. Miotay. (1991). Fruit and seed set in the cucumber (*Cucumis sativus* L.) in relation to pollen tube growth, sex type, and parthenocarpy. *Gartenbau.* 56: 46-49.

Nijs, A. P. M. den and D. L. Visser. (1980). Induction of male flowering in gynoecious cucumbers (*Cucumis sativus* L.) by silver ions. *Euphytica* 29: 273-280.

Nishi. (1955). The commercial seed production of F_1 varieties of vegetables in Japan. *Rpt. 14th Int. Hort. Congr.* 1: 468-478.

Nishi. (1967). F_1 seed production in Japan. *Proc. XVII Internatl. Hort. Congr* 3: 231-257.

Nugent, P. E. and J. C. Hoffman. (1981). Natural cross pollination in four andromonecious seedling marker lines of muskmelon. *HortScience* 16: 73-74.

Om, Y. H., D. G. Oh, and K. H. Hong. (1992). Evaluation of heterosis and combining ability for several major characters in oriental melon. *Res. Rept. Rural Dev. Adm. (Suweon):* 29: 74-76.

Pal, B. P. and H. Singh. (1946). Studies in hybrid vigour. II. Notes on the manifestation of hybrid vigour in the brinjal and bitter gourd. *Ind. J. Genet. Pl. Breed.* 61: 19-33.

Passmore, S. F. (1934). Hybrid vigour in reciprocal crosses in *Cucurbita pepo. Ann. Bot.* 48: 1029-1030.

Pearson, O. H. (1983). Heterosis in vegetable crops. In: *Heterosis*, ed, R. Frankel. Springer-Verlag, Berlin. pp. 138-188.

Peterson, C. E. (1960). A gynoecious inbred line of cucumber. *Quart. Bull. Mich. Agr. Expt. Sta.* 43: 40-42.

Peterson, C. E. and L. D. Anhder. (1960). Induction of staminate flowers on gynoecious cucumbers with gibberellin A_3. *Science* 131: 1673-1674.

Peterson, C. E. and D. J. DeZeeuw. (1963). The hybrid pickling cucumber, Spartan Dawn. *Quart. Bull. Mich. Agr. Expt. Sta.* 46: 267-273.

Peterson, C. E., K. E. Owens, and P. R. Rowe. (1983). Wisconsin 998 muskmelon germplasm. *HortScience* 18: 116.

Peterson, C. E. and J. L. Weigle. (1957). A new method for producing hybrid cucumber seed. *Quart. Bull. Mich. Agr. Expt. Sta.* 40: 960-965.

Peto, H. B. (1950). Pollination of cucumbers, watermelons and cantaloupes. *Iowa State Apiarist Rpt.*: 79-87.

Pike, L. M. and W. A. Mulkey. (1971). Use of hermaphrodite cucumber lines in development of gynoecious hybrids. *HortScience* 6: 339-340.

Pike, L. M. and C. E. Peterson. (1969). Gibberellin A_4/A_7 for induction of staminate flowers on the gynoecious cucumber (*Cucumis sativus* L.). *Euphytica* 18: 106-109.

Pitchaimuthu, M. and P. S. Sirohi. (1997). Genetic analysis of fruit characters in bottle gourd (*Lagenaria siceraria* (Mol.) Standl.). *J. Genet Breed.* 51: 33-37.

Porter, D. R. (1933). Watermelon breeding. *Hilgardia* 7: 585-624.

Raj, N. M., K. P. Prasanna, and K. V. Peter. (1992a). Bitter gourd *Momordica* spp. In: *Genetic Improvement of Vegetable Crops,* eds. G. Kalloo and B. O. Bergh. Pergamon Press, Oxford. pp. 239-246.

Raj, N. M., K. P. Prasanna, and K. V. Peter. (1992b). Snake gourd *Trichosanthes anguina* L. In: *Genetic Improvement of Vegetable Crops*, eds. G. Kalloo and B. O. Bergh. Pergamon Press, Oxford. pp. 259-264.

Randhawa, K. S. and M. J. Singh. (1990). Assessment of combining ability, heterosis and genetic variance for fruit quality characters in muskmelon (*Cucumis melo* L.). *Indian J. Genet.* 50: 127-130.

Rhodes, B. B., W. C. Adamson, and W. C. Bridges. (1987). Outcrossing in watermelons. *Cucurbit Genet. Coop. Rept.* 10: 66-68.

Robinson, R. W. , S. Shannon, and M. D. de la Guardia. (1969). Regulation of sex expression in the cucumber. *BioScience* 19: 141-142.

Robinson, R. W. (1992). Genetic resistance in the Cucurbitaceae to insects and spider mites. *Plant Breed. Rev.* 10: 309-360.

Robinson, R. W., H. M. Munger, T. W. Whitaker, and G. W. Bohn. (1976). Genes of the Cucurbitaceae. *HortScience* 11: 554-568.

Robinson, R. W. and D. S. Decker-Walters. (1997). *Cucurbits*. CAB Internatl. Wallingford, UK. 226 pp.

Robinson, R. W., M. A. Boettger, and J. W. Shail. (1978). Gynoecious sex expression in *Cucurbita* resulting from an interspecific cross. *Cucurbit Genetics Coop. Rpt.* 1: 31-32.

Robinson, R. W., T. W. Whitaker, and G. W. Bohn. (1970). Promotion of pistillate flowering in *Cucurbita* by 2-chloroethylphosphonic acid. *Euphytica* 19:180-183.

Rosa, J. T. (1924). Fruiting habit and pollination of cantaloupe. *Proc. Amer. Soc. Hort. Sci.* 21: 51-57.

Rosa, J. T. (1927). Results of inbreeding melons. *Proc. Amer. Soc. Hort. Sci.* 24: 79-84.

Rubino, D. B. and T. C. Wehner. (1986). Effect of inbreeding on horticultural performance of lines developed from an open-pollinated pickling cucumber *Cucumis sativus* population. *Euphytica* 35: 459-464.

Rudich, J., N. Kedar, and A. H. Halevy. (1970). Changed sex expression and possibilities for F_1 hybrid seed production in some cucurbits by application of Ethrel and Alar (B-995). *Euphytica* 19: 47-53.

Sageret, A. (1826). Considerations sur la production des hybrids des variantes et des varietés en general, et sur celles des Cucurbitaceés en partuculier. *Ann. Sci. Nat. Prem.* Sér. 8: 294-314.

Scheerens, J. C., H. M. Scheerens, A. E. Ralowicz, T. L. McGriff, M. J. Kopplin, and A. C. Gathman. (1988). Staminate floral induction on gynoecious buffalo gourd following application of AVG. *HortScience* 23: 138-140.

Schultheis, J. R., J. T. Ambrose, S. B. Bambara, and W. A. Mangum. (1994). Selective bee attractants did not improve cucumber and watermelon yield. *HortScience* 29: 155-158.

Schuster, W., M. R. Haghdadi, and J. Michael. (1974a). Inzucht und Heterosis bei Olkürbis (*Cucurbita pepo* L.), I. Inzuchtwirkung. *Z. Pflanzenzüchtg.* 73: 112-124.

Schuster, W., M. R. Haghdadi, and J. Michael. (1974b). Inzucht und Heterosis bei Olkürbis (*Cucurbita pepo* L.), II. Bastardwüchsigkeit. *Z. Pflanzenzüchtg.* 73: 233-248.

Scott, G. W. (1933). Inbreeding studies with *Cucumis melo*. *Proc. Amer. Soc. Hort. Sci.* 29: 485.

Scott, G. W. (1934). Observations on some inbred lines of bush types of *C. pepo*. *Proc. Amer. Soc. Hort. Sci.* 32: 480.

Scott, J. W. and L. R. Baker. (1976). Sex expression of single and 3-way cross cucumber hybrids with androecious pollinators. *HortScience* 11: 243-245.

Seshadri, V. S. (1986). Cucurbits. In: *Vegetable Crops in India*, eds. T. K. Bose and M. G. Som. Naya Prokash, Calcutta.

Shannon, S. and R. W. Robinson. (1979). The use of ethephon to regulate sex expression of summer squash for hybrid seed production. *J. Amer. Soc. Hort. Sci.* 104: 674-677.

She, M., H. Zhou, and Y. Li. (1996). A new hybrid of white gourd–Qingza 1. *China Vegetables* 3: 7-10.

Shifriss, O. (1945). Male sterilies and albino seedlings in cucurbits. *J. Hered.* 36: 47-52.

Shifriss, O. (1987). Synthesis of genetic females and their use in hybrid seed production. U. S. Patent No. 4,686,319.

Shifriss, O. (1988). On the emergence of B cultivars in squash. *HortScience* 23: 238.

Shifriss, O. and W. L. George Jr. (1964). Sensitivity of female inbreds of Cucumis sativus to sex reversion by gibberellin. *Science* 143:1452-1453.

Shimotsuma, M. (1959). Cytogenetical studies in the genus *Citrullus*, II. Intra- and interspecific hybrids obtained from all possible cross combinations between diploid and tetraploid *C. colocynthis* Schrad. and *C. vulgaris* Schrad. *Seiken Ziho* 10: 37-48.

Shimotsuma, M. (1962). Studies on triploid seed production in watermelons. *Jap. J. Breed.* 12: 124-130.

Singh, J. P., H. S. Gill, and K. S. Ahluwalia. (1970). Studies in hybrid vigour in cucumbers (*Cucumis sativus* L.). *Indian J. Hort.* 27: 36-38.

Singletary, C. C. and E. L. Moore. (1965). Hybrid watermelon seed production. *Mississippi Farm Res.* 28 (6): 5.

Sinnott, E. W. and G. B. Durham. (1922). Inheritance in the summer squash. *J. Hered.* 13: 177-186.

Skeels, F. E. (1887). Cross fertilization of cucurbits. *Agr. Sci.* 1: 228-229.

Solanki, S. S., J. N. Seth, and S. D. Lal. (1987a). Heterosis and inbreeding depression in cucumber *Cucumis sativus* L. II. *Prog. Hort.* 20: 235-239.

Solanki, S. S., J. N. Seth, and S. D. Lal. (1987b). Heterosis and inbreeding depression in cucumber *Cucumis sativus* L. III. *Prog. Hort.* 20: 248-252.

Solanki, S. S., J. N. Seth, and S. D. Lal. (1988). Heterosis and inbreeding depression in cucumber *Cucumis sativus* L. I. *Prog. Hort.* 20: 15-19.

Staub, J. E., B. Balgooyen, and G. E. Tolla. (1986). Quality and yield of cucumber hybrids using gynoecious and bisexual parents. *HortScience* 21: 510-512.

Stino, K. R., W. A. Warid, and M. A. Abdelfattah. (1963). Some examples of heterosis in the melon, *Cucumis melo* L. *Cairo Univ. Bull.* 224. 11 pp.

Takahashi, O. (1987). Utilization and seed production of hybrid vegetable varieties in Japan. In: *Hybrid Seed Production of Selected Cereal Oil and Vegetable Crops*, eds. W. P. Feistritzer and A. G. Kelly. FAO Plant Production and Protection Paper 82: 313-328.

Tasdighi, M. and L. R. Baker. (1981). Comparison of single and three-way crosses of pickling cucumber hybrids for femaleness and yield by once-over harvest. *J. Amer. Soc. Hort. Sci.* 106: 370-373.

Tatlioglu, T. (1993). Cucumber *Cucumis sativus* L. In: *Genetic Improvement of Vegetable Crops*, eds. G. Kalloo and B. O. Bergh. Pergamon Press, Oxford. pp 197-234.

Tolla, G. E. and C. E. Peterson. (1979). Comparison of gibberellin A_4/A_7 and silver nitrate for induction of staminate flowers in a gynoecious cucumber line. *HortScience* 14: 542-544.

Wall, J. R. (1960). Use of marker genes in producing triploid watermelons. *Proc. Amer. Soc. Hort. Sci.* 76: 577-581.

Wall, J. R. (1967). Correlated inheritance of sex expression and fruit shape in *Cucumis. Euphytica* 16: 69-76.

Watts, V. M. (1962). A marked male-sterile mutant in watermelon. *Proc. Amer. Soc. Hort. Sci.* 81: 498-505.

Wehner, T. C. and S. F. Jenkins Jr. (1985). Rate of natural outcrossing in monoecious cucumbers. *HortScience* 20: 211-213.

Wehner, T. C. and C. H. Miller. (1985). Effect of gynoecious expression on yield and earliness of a fresh-market cucumber hybrid. *J. Amer. Soc. Hort. Sci.* 110: 464-466.

Wehner, T. C. and R. W. Robinson. (1991). A brief history of the development of cucumber cultivars in the U.S. *Cucurbit Genet. Coop. Rept.* 14: 1-4.

Whelan, E. D. P. (1974). Linkage of male sterility, glabrate seedling and determinate plant habit in cucumber. *HortScience* 9: 576-577.

Whitaker, T. W. and G. W. Bohn. (1952). Natural cross pollination in muskmelon. *Proc. Amer. Soc. Hort. Sci.* 60: 391-396.

Whitaker, T. W. and G. N. Davis. (1962). *Cucurbits Botany, Cultivation, and Utilization.* Interscience Publ., NY. 249 pp.

Whitaker, T. W. and D. E. Pryor. (1946). Effect of plant growth regulators on the set of fruit from hand-pollinated flowers in *Cucumis melo* L. *Proc. Amer. Soc. Hort. Sci.* 48: 417-422.

Whitaker, T. W. and R. W. Robinson. (1986). Squash breeding. In: *Breeding Vegetable Crops*, ed. M. J. Bassett. Avi Publ. Co., Westport, CT. pp. 209-242.

Wittwer, S. H. (1974). Growth regulants in agriculture. *Plant Growth Reglulator Working Group Bull.* 2 (1): 5-8.

Wolf, E. A. and J. D. Hartman. (1942). Plant- and fruit-pruning as a means of increasing fruit set in muskmelon breeding. *Proc. Amer. Soc. Hort. Sci.* 40: 415-420.

Yamashita, T. (1973). Current utilization of F_1 hybrids for vegetable production in Japan. *Jap. Agr. Res. Quarterly* 7: 195-201.

Yu, C.-H. (1982). Breeding for hybrid varieties and production of F_1 seeds in vegetables. *Proc. Symp. Plant Breed.*, Taichung, Taiwan: 211-217.

Zhang, Q., A. C. Gabert, and J. R. Baggett. (1994). Characterizing a cucumber pollen sterile mutant: inheritance, allelism, and response to chemical and environmental factors. *J. Amer. Soc. Hort. Sci.* 119: 804-807.

Zhang, X. and B. Rhodes. (1993). Watermelon variety improvement in China. *Cucurbit Genet. Coop Rept.* 15: 76-79.

Zhang, X. P., B. B. Rhodes, W. V. Baird, H. T. Skorupska, and W. C. Bridges. (1996). Development of genic male-sterile watermelon lines with delayed-green seedling marker. *HortScience* 31: 123-126.

Hybrid Seed Production in *Capsicum*

Terry G. Berke

SUMMARY. While ~20 *Capsicum* species are recognized, *Capsicum annuum* is the predominant species cultivated, encompassing both hot and sweet peppers. Heterosis has been documented in hot and sweet peppers, and hybrids are gaining increasing popularity among farmers throughout the world. Producing high-quality hybrid pepper seeds requires careful management of the parental lines, skilled labor to make the cross-pollinations, and proper processing of the resulting seeds. The use of genic and cytoplasmic-genic male sterility is increasing in order to decrease the cost of hybrid seed production. Molecular markers are increasingly being used to check the purity of hybrid seeds. Most hybrid seed production occurs in countries with cheap, skilled labor, such as China, India, and Thailand. *[Article copies available for a fee from The Haworth Document Delivery Service: 1-800-342-9678. E-mail address: getinfo@haworthpressinc.com <Website: http://www.haworthpressinc.com>]*

KEYWORDS. *Capsicum*, cross-pollination, hybrid, male sterility, pepper, production, seed

Pepper (*Capsicum* spp.) is a dicotyledonous herb with erect, compact, or sometimes prostrate growth habit that is adapted to many climates (Andrews, 1995). It is grown for its nonpulpy berry (commonly called a fruit or pod), which may be eaten flesh, cooked, or as a

Terry G. Berke is Associate Scientist, Asian Vegetable Research and Development Center, Box 42, Shanhua, Tainan, 741, Taiwan ROC (E-mail: terry@netra.avrdc.org.tw).

[Haworth co-indexing entry note]: "Hybrid Seed Production in *Capsicum*." Berke, Terry G. Co-published simultaneously in *Journal of New Seeds* (Food Products Press, an imprint of The Haworth Press, Inc.) Vol. 1, No. 3/4, 1999, pp. 49-67; and: *Hybrid Seed Production in Vegetables: Rationale and Methods in Selected Crops* (ed: Amarjit S. Basra) Food Products Press, an imprint of The Haworth Press, Inc., 2000, pp. 49-67. Single or multiple copies of this article are available for a fee from The Haworth Document Delivery Service [1-800-342-9678, 9:00 a.m. - 5:00 p.m. (EST). E-mail address: getinfo@haworthpressinc.com].

dried powder (Poulos, 1993). Pepper originated in the New World, and peppers share the distinction of being the first plants cultivated in the New World with beans (*Phaseolus* spp.), maize (*Zea mays* L.), and cucurbits (Cucurbitaceae) (Heiser, 1973). They are also one of the first spices found to have been used by humans anywhere in the world. Widespread geographic distribution of *C. annuum* and *C. frutescens* occurred in the 16th century via Spanish and Portuguese traders to all continents, whereas the other species are little distributed outside South America (Andrews, 1995).

Peppers belong to the family Solanaceae, which also includes the widely cultivated genera *Lycopersicon* (tomato) and *Solanum* (potato and eggplant). Within the genus *Capsicum*, five species are commonly recognized as domesticated: *C. annuum, C. baccatum, C. chinense, C. frutescens,* and *C. pubescens,* while approximately 20 wild species are documented including *C. chacoense, C. exirnium,* and *C. praeterrnissum.* Wild forms of all the domesticated species exist except for *C. pubescens,* which is known only in cultivation (Smith and Heiser, 1957). The domestication of the different *Capsicum* species is thought to have taken place independently in different areas of the New World. *Capsicum annuum* was Mesoamerican, *C. baccatum* was Andean, *C. chinense* and *C. fiutescens* were Amazonian, and *C. pubescens* was mid-elevational (2,000-2,500 m) Andean (Andrews, 1995). The *Capsicum taxa* do not cross easily in all combinations, although almost all species have the somatic chromosome number of 2n = 24. Detailed diagrams of species cross-compatibility may be found in Creenleaf (1986) and Zijlstra, 3 Purimahua, and Lindhout (1991). No successful hybridizations between *Capsicum* and other species in the Solanaceae family have been reported.

Capsicum germplasm as seed is available from a number of sources (Berke and Engle, 1997). The Asian Vegetable Research and Development Center (AVRDC) in Shanhua, Taiwan has the world's largest collection, with 7,273 accessions (AVRDC, 1999). Other large collections are at the Southern Regional Plant Introduction Station in Griffin, Georgia with 4,682 accessions (USDA-ARS, 1999), and the Instituto Nacional de Investigaciones Forestales y Agropecuarias (INIFAP) in Celaya, Guanajuato, Mexico, with 3,590 accessions (Bettencourt and Konopka, 1990). Many smaller collections are maintained at various institutes worldwide (Bettencourt and Konopka, 1990). Information regarding the genetics of *Capsicum* is published through peer-reviewed journals, the EUCARPIA Capsicum and Eggplant Newsletter, the biennial National Pepper Conference proceedings in the U.S., and the triennial EUCARPIA meeting on the genetics and breeding of *Capsicum* and eggplant proceedings in Europe.

PLANT CULTURE

Field Cultivation

Peppers are normally started in a seedling tray or bed in a sterile medium such as peat moss. Peppers are slow to germinate, and their germination rate is dependent on the temperature. Optimum germination occurs between 22 and 30°C (Maynard and Hochmuth, 1997). At these temperatures, peppers will usually germinate in 8-10 days. Seed priming may be used to obtain earlier and more uniform seed germination. Peppers develop best in a warm, moderate climate that is not too hot to set fruit. Daytime temperatures of 25°C and nighttime temperatures of 18°C are ideal for growing peppers. The soil fertility requirements of peppers are relatively high, but the amount of fertilizer applied will depend on the soil type, soil fertility, previous crop, and plant density. Peppers grow best in a soil that is moist, but not wet. Wet soils frequently produce unhealthy plants suffering from root rot (caused by *Phytophthora capsici*).

Some of the plants in a field nursery may be grown with net or stake supports to facilitate observation, hybridization, selection, classification, and maximum yields of high quality seeds. Plant population density varies depending on the size of the plants but 2-3 plants m^{-2} are common, either in single rows spaced 1 m apart or in double rows spaced 0.5 m apart on beds. Double-row, unsupported plots planted at the rate of 2 plants m^{-1} may be used for evaluation of lines and for seed increase. Large-scale increases of advanced lines may be conducted in isolated fields or in nethouses to prevent out-crossing by bees.

Greenhouse and Screenhouse

Greenhouses and screenhouses are routinely used in pepper breeding. They are a convenient way to obtain controlled self-pollination in pepper without having to cover plants in the open field. Plants may be grown in a potting soil mixture, an inert medium like sand, or hydro-ponically. Nutrients should be supplied as needed. Moisture may be supplied by hand watering or by any of various automatic devices such as trickle irrigation tubes. The medium should not become water-logged.

Common greenhouse temperatures are a constant 20°C, or 25°C day/18°C night. Peppers are classified as photoperiod-insensitive, so supplemental lighting is not needed under short-day conditions to induce flowering, but it is beneficial in promoting growth in temperate regions during the winter months.

FLORAL CHARACTERISTICS

The pepper usually is considered to be self-pollinated; however, cross-pollination can be quite extensive with some genotypes and environments. Reported percentages of natural cross-pollination (NCP) range from 0-91% (Odland and Porter, 1941; Cotter, 1980; Tanksley, 1984; Todorov and Csillery, 1990; and Kanwar, 1995). The variation in percent NCP is a function primarily of bee activity and heterostyly of the flowers. Wind is not generally considered a factor in NCP. In climates with insect pollinators present, controlled self-pollinations should be made to ensure genetic purity of selected plants. Covering the plants with nets to prevent insects, primarily bees (*Apis* spp.), from visiting the plants will help ensure self-pollination occurs. Pollen placed on the stigma remains inactive for some time. Time for pollen germination ranges from 6-42 h after pollination (Cochran, 1938; Hirose, 1965; and Dumas De Vaulx and Pitrat, 1977). Time from pollination to fertilization was reported to be approximately 72 hours (Dumas De Vaulx and Pitrat, 1977).

The flowers of *C. annuum* are usually borne singly and terminal, although a fasciculate gene, *fa*, conditioning multiple flowers per node has been reported (Lippert, Gergh, and Smith, 1965). *Capsicum chinense* and *C. frutescens* have from 1-5 flowers per node and the inheritance of this trait in *C. chinense* has been reported to be polygenic (Subramanya, 1983). The pedicel length varies greatly depending on genotype, ranging from 3-8 cm (Poulos, 1993).

Floral structure of the pepper is characteristic of plants belonging to the family Solanaceae, a perfect flower containing a pistil, stamens, petals, and fused sepals forming a calyx (Figure 1). The green calyx is

FIGURE 1a. *Capsicum* flower structure.

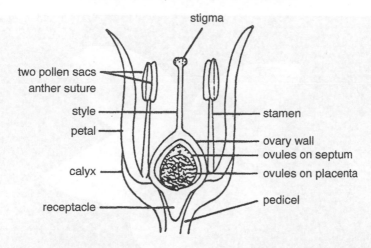

FIGURE 1b. *Capsicum* flower (cutaway side view).

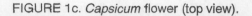

FIGURE 1c. *Capsicum* flower (top view).

cup-shaped, persistent, and enlarging in the fruit, usually with five conspicuous teeth. The corolla is campanulate to rotate with 5-7 lobes, 8-15 mm in diameter and usually white in *C. annuum*, although purple is sometimes observed. *Capsicum frutescens* is distinguished by its pale greenish white corolla, *C. pubescens* always has black seeds and a purple corolla, and *C. baccatum* is distinguished by its greenish-yellow corolla throat spots.

Capsicum annuum usually has 5-7 stamens with pale-blue to purple anthers. A series of anthocyanin-less genes, *al1-al5*, conditioning yellow anthers have been reported (Daskalov and Poulos, 1994). *Capsicum chinense*, *C. fiutescens*, and *C. pubescens* typically have blue-purple anthers, while *C. baccatum* has yellow anthers. The pistil is composed of an ovary with a longitudinal diameter of 2-5 mm containing 2-4 carpels (locules), a style 3.5-6.5 mm long, and a capitate, lobed, papillate stigma that has a mean diameter slightly greater than that of the style. The relative position of the stigma and the anthers may vary considerably according to the genotype, but they can generally be classified by the length of their styles into short-styled, medium-styled, and long-styled (Quagliotti, 1979).

The receptivity of the stigma varies with temperatures during and after anthesis. It is highest on the day of anthesis when the anthers are fully developed but are still indehiscent and the corolla is ready to open (Cochran and Dempsey, 1966). In non-emasculated flowers, receptivity lasts for 4-7 days, and in emasculated flowers it lasts for 5-9 days (Markus, 1969). Pollen can fertilize the egg from the time the anthers burst until 24-48 h later (at 20-30°C). Pollen can be stored at 0°C for 5-6 days, and if the pollen is dried before storage it can be preserved at 0°C for fairly long periods (Quagliotti, 1979).

The first node of the plant usually bears the first flower, and it may undergo anthesis as much as 7-10 days before the flowers at the second node. Nodal development is dichotomous in most genotypes, although genotypes with a high degree of prebifurcation (side branching below the first node) have been reported (Shifriss and Yehuda, 1977). Anthesis proceeds sequentially upward from node to node. A determinate plant type produces a limited number of nodes, but the number of nodes depends on the genotype and the environment. An indeterminate plant type keeps growing and producing nodes until frost, disease, or some other mechanism halts its growth. Earliness of flowering is a function primarily of the genotype although some interaction with temperature does occur.

ARTIFICIAL HYBRIDIZATION AND SELF-POLLINATION

Equipment

The equipment needed for artificial hybridization includes a pair of forceps with sharp tips and rings or tags to mark cross-pollinated fruits (Figure 2). For commercial hybrid seed production, you may need a small plastic container with an airtight lid for collecting and storing pollen, an airtight box containing silica gel or $CaCO_3$ for drying anthers, a fine brush, wax paper, and fine mesh cloth for pollen extraction (Figure 2). Alternatively, a vibrator with a little collector made of glass can be used to collect pollen. A pollination ring may be used to carry pollen when large numbers of pollinations are done using pollen from the same male parent. This is a small ring with a hollow cylinder

FIGURE 2. Pollination equipment.

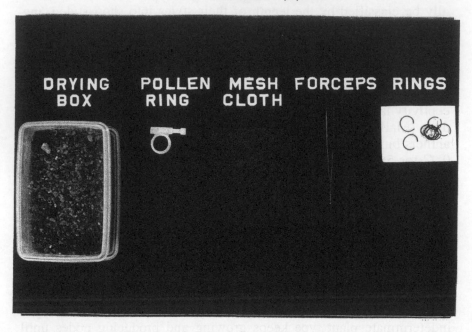

attached to it that is plugged with a smaller solid cylinder that can slide up and down inside it to control the volume (Figure 2).

Preparation of the Female

Female plants should be vigorous, so plants should be managed carefully with optimum weed control, fertilization, staking, and spacing to obtain vigorous plants. Female plant population is usually 30,000-40,000 plants ha^{-1}; the male plant population density can be higher than the female. Plants should be grown in a relatively cool (maximum temperature < 25 and < 30°C for sweet and chilli peppers, respectively), dry environment to minimize disease and insect infestations. The leaves and flowers should not become wet, so water should be provided by drip or furrow irrigation. Irrigation should be in the late afternoon or at night. Lighter, more sandy soils produce fruits with fewer black seeds and brighter seed color. Lime (CaCO$_3$) and MgSO$_4$ should be applied to prevent blossom-end rot in sweet peppers. Before flowering, the amount of sidedressed N should be reduced. During pollination, KH_2PO_4 as a source of P and K should be applied weekly as a foliar fertilizer to stimulate flowering and fruit set.

First, prepare the female plants for pollination by removing all self-pollinated fruits and open flowers, leaving only closed flowers. Greatest success in hybridization was reported at nodes 2-4 on a plant (Rylski, 1984). Another researcher (Lillian Wong, 1996, pers. comm.) reports that nodes 4-5 give the greatest success in sweet peppers and nodes 5-6 give the greatest success in chilli peppers. The flower to be emasculated should have developed to the stage just prior to anther dehiscence (Marffitina, 1975; Aleemullah, Haigh, and Holford, 1996). At this stage the anthers can be carefully removed. Care must be taken to avoid damaging the pistil. Younger flower buds are more difficult to emasculate and are more easily damaged. Older flower buds will already be self-fertilized. If two or more flowers are present at a node, all but one should be snipped off to reduce competition for nutrients and promote fruit growth. For openfield hybrid seed production in sweet bell peppers, 6-8 fruits per plant are considered the optimum number for maximum seed set and seed vigor. For elongate bell types, 7-9 fruits per plant are considered the optimum number. For chilli peppers, optimum fruit number depends on fruit size. Generally, 15-20 fruits per plant are considered the optimum number. In Korea they usually harvest 50-70 fruits per plant (Il-Woong Yu, 1996, pers. comm.). Hybrid seed production in greenhouses can last much longer, from 6-10 months, with 2-3 times the number of fruits per plant and multiple fruit harvests.

Flowers may be emasculated in the early morning or late afternoon. Begin emasculation by removing the petals from the flower, then carefully remove the anthers without damaging the pistil (Figure 3). Once the petals are removed bees usually do not visit the flowers so there is no need to cover the emasculated flower (wind-carried pollen is not considered to be a cause of cross-pollination in pepper). If pollinations are performed where bees are not present, there is no need to remove the petals. Mark emasculated flowers with a clip for easy location the next day. Stigma receptivity to pollen is highest on the day of anthesis, when fully developed anthers are still closed.

Pollination

Pollen for crossing can be obtained at any time of day, with the most abundant and viable pollen coming from freshly dehisced anthers. The

FIGURE 3a. Clasping the anthers.

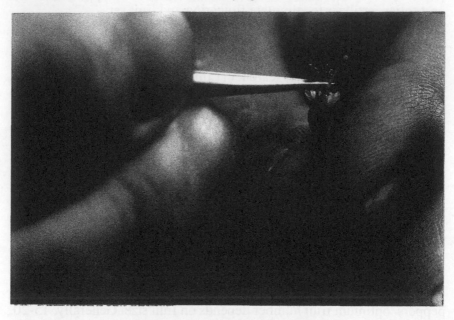

FIGURE 3b. Removing the anthers.

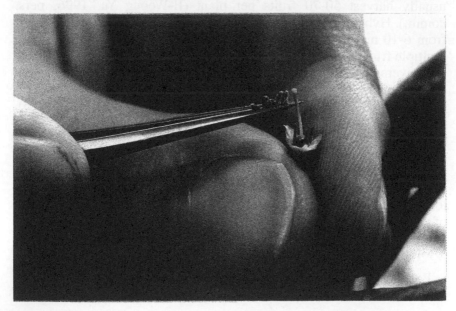

optimal temperature for pollen germination is 20-25°C (Quagliotti, 1979). If the temperature is above 30°C, the viability of the pollen is considerably reduced (Cochran, 1938). High temperatures hinder the formation of pollen. If relatively few crosses are made inside a greenhouse or growth chamber, there is no need to collect pollen. Freshly dehisced anthers can be picked from the male parent and used to make the pollination.

For commercial hybrid seed production, pollen must be collected from the male parent. If other peppers are growing within 200 m, flowers should be harvested from the male parent when they are still closed, just prior to opening, to prevent contamination from outside pollen. If there are no other peppers in the area, flowers can be harvested when they have newly opened. Anthers should be removed from the flowers with forceps and placed in wax paper inside a drying box with silica gel or $CaCO_3$ overnight. Alternatively, anthers can be placed 50-60 cm beneath an incandescent lamp for several hours. Dried anthers should be placed in a fine mesh bag with a small coin or rock inside. The bag should be placed inside an airtight plastic container and shaken vigorously to separate the pollen from the anthers. The pollen will fall outside the bag where it can be collected and the remaining contents of the bag can be discarded. Dried pollen can be stored at room temperature (20-25°C) for approximately 24-48 h, and for up to a week at 0°C (Quagliotti, 1979). If pollen is stored in sealed vials or capsules containing silica gel or calcium sulfate, it will remain viable for one month at 4°C (Poulos, 1993).

A pollination ring is generally used to carry pollen in the field while pollinating. It is filled with pollen and may be worn on either hand. Emasculated flowers on female plants are located by the clips marking them and the stigma of the emasculated flower is dipped in the pollen (Figure 4). Alternatively, a fine brush can be used to apply pollen. The clip is removed and a tag or ring is placed around the pedicel or the node to mark the cross-pollinated flower. This process is repeated until all the emasculated flowers are pollinated or all the pollen is gone. The pollination success rate is generally 60-80% (Lillian Wong, 1996, pers. comm.). One person can emasculate and pollinate approximately 400 plants per day. For maximum fruit set, the best time of day to pollinate is after the dew dries in the morning until 10 a.m. or after 3 p.m. in the afternoon when temperatures are cooler and relative humidity and turgor of the plant are higher. Each day between 10 a.m. and 3

FIGURE 4. Dipping the stigma in the pollen.

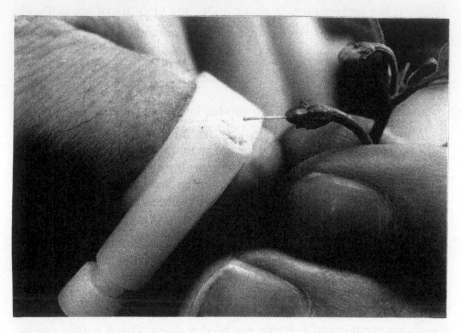

p.m. flowers at the proper stage of development on the female plants should be emasculated. Pollinations should be done on flowers emasculated the previous day until the optimum number of cross-pollinated fruits is obtained on each female plant. Number of seeds per fruit will vary with genotype and environment.

The night temperature has the most important effect on fertilization and seed set. If the night temperature is lower than 15°C, sweet peppers will have few seeds per fruit. Night temperatures that are lower than 10°C prevent fertilization and promote parthenocarpic fruit set (Rylski, 1984). Night temperatures of 17-20°C promote optimum fertilization and fruit set, whereas night temperatures above 21°C promote blossom drop (Cochran, 1936; Rylski and Spigelman, 1982). Optimum day temperatures range from 25-28°C, but high day temperatures (32°C) do not increase blossom drop as long as night temperatures are optimum (Rylski, 1984). Very low humidity also prevents fertilization and promotes blossom drop.

FACTORS AFFECTING EFFICIENCY

Pollen availability is crucial. Ratios of 5:1 and 10:1 female:male plants are commonly used for sweet and chilli pepper hybrid seed production, respectively. To further ensure an adequate supply of pollen at pollination time, male:female flowering is synchronized by sowing the male line one week earlier than the female line if they flower at the same time. If they flower at different times then the sowing date is adjusted accordingly. Alternatively, you can use multiple sowing dates for the male line. Since peppers generally blossom over a 3-4 week period no further efforts are usually needed to ensure an adequate supply of pollen.

Hybrid seed companies usually conduct grow-outs of F_1 plants to check the purity of the hybrid seed. Various genetic markers such as *all*, L^1, or rugose leaf (*ru1*), can be used in the F_1 to check the purity. If the male line has the dominant allele L^1 for TMV resistance and the female line is L^+, self-pollinated plants will be susceptible to TMV and hybrid plants will be resistant. If the female line carries a recessive marker such as *ru1*, self-pollinated plants will have rugose leaves and hybrid plants will have normal leaves. If the male line carries a dominant marker for purple internodes, self-pollinated seedlings can be detected as early as 21 days after sowing (Lillian Wong, 1996, pers. comm.). Alternatively, isozymes or molecular markers such as randomly amplified polymorphic DNA (RAPDs) can be used to check the purity of hybrid seed.

MALE STERILITY

Hybrid seed production is labor-intensive, particularly emasculation of flowers on female plants. Male sterility can be used to reduce the time needed to prepare the female flowers for pollination and thus reduce the amount of labor needed. Male sterility also increases the purity of the F_1 seed since no self-pollination takes place.

Cytoplasmic-genic male sterility (cgms) was first reported by Peterson (1958) in an introduction of *C. annuum* from India (PI 164835). Most authors have reported that a single nuclear gene, designated *rf1*, interacts with S cytoplasm to produce sterility, and the restorer allele *Rf1* restores fertility (Shifriss, 1997). Plants with N cytoplasm are fertile regardless of whether they have the *Rf1* or *rf1* allele. Peterson

(1958) and Novak, Betlach, and Dubovsky (1971) reported that a few genotypes apparently carried a second nuclear gene, designated *rf2*, which was needed in addition to *rf1* in order to produce sterility in S cytoplasm.

Many authors have reported that cgms can be unstable under different environmental conditions, depending on the genotype. In Korea, hybrid seed companies such as Choong Ang and Hung Nong have been using cgms for several years for hybrid seed production, and their A lines are stable under diverse environmental conditions (Il Woong Yu, 1998, pers. comm.). Yu (1990) in a study of 270 lines, found 152 to be stable maintainers, 66 to be stable restorers, and 52 to be unstable, with large deviations in sterility in different environments. Out of 39 large-fruited sweet peppers, 20 were stable maintainers, and 18 were unstable genotypes. Although different sources of male sterile cytoplasm have been reported in *C. annuum*, they all appear to be identical (Shifriss, 1997).

Cytoplasmic male sterility has also been reported by interspecific hybridization of *C. baccatum* × *C. annuum*. No adequate restorers have been found for this system. Also, *C. chacoense* × *C. annuum* has been reported to give male sterility. No restorers or maintainers have been found.

Several stable, recessive nuclear genes for male sterility are available and being used in several breeding programs around the world. At least 12 genes have been reported, obtained mostly by mutagenesis (Shifriss, 1997). Few allelism tests have been performed, although it was shown that the *ms509* gene from France is allelic to the *msk* gene from Korea but not to *ms3* or *ms705* (Yu, 1990). Some of them are linked with marker traits that can help in early identification of the msms individuals (Murty and Lakshmi, 1979; Meshram and Narkhede, 1982; and Pathak, Singh, and Deshpande, 1983). When marker genes are not available, segregating female plants may be planted at 2-3 times the normal density; *msms* plants are identified at anthesis, and fertile *Msms* plants are discarded. Kaul (1988) and Shifriss (1997) have published extensive reviews of male sterility in peppers.

SEED DEVELOPMENT, HARVEST, AND STORAGE

The developing fruit of successfully fertilized flowers will begin to extend past the pedicel in 4-6 days. Non-fertilized flowers will occasionally set parthenocarpic fruits, otherwise they drop off within 3-4 days alter pollination. Parthenocarpic fruits are usually smaller and somewhat deformed compared to seed-bearing fruits (Rylski, 1984). Fruit elongation follows a simple sigmoidal curve (Rylski, 1984). Seed maturation takes 25-85 days, depending on the genotype and the environment in which it is grown. The stage for maximum seed germination approximates the stage of fruit ripeness as indicated by color change in the fruit wall, usually green (immature) to red or yellow (mature). Fruits should be harvested at the fully mature stage for maximum germination rate (Edwards and Sundstrom, 1987; Demir and Ellis, 1992; Sanchez et al., 1993; Cavero, Gil Ortega, and Zaragoza, 1995). However, this depends on the temperature during fruit development (Ken Owens, 1996, pers. comm.). Under cold conditions, the seeds may mature before the fruit turns color. Under hot conditions, the seeds may not mature before the fruit turns color, and the fruits should be left on the plant longer. If fruits are left on the plant too long, germination rate may decrease, especially in sweet peppers.

Seeds from hand-pollinated fruits may be extracted by hand, or a modified meat grinder may be used to macerate the fruits (Tay, 1991). The seeds and macerated fruit pulp are placed in a container filled with water. Seeds with good germination and good seedling vigor are generally bright yellow and plump, and they sink in water. Seeds with poor or no germination are generally dull yellow, brown, or black, and they float in water along with the macerated fruit pulp. Several rinses may be performed until only plump seeds remain on the bottom of the container. They are then removed from the water and placed in mesh bags. Excess water may be extracted by spinning them in a centrifuge or similar device. They should then be dried in a dry, shaded area with good ventilation or in a seed dryer at 20°C and 40% relative humidity (RH). Sweet pepper seed extraction is usually done by hand because water turns the seed color from bright yellow to dull yellow or tan. Seed color is not important if the seeds will be treated with a colored fungicide or other chemical.

Several pepper pathogens are seed-borne, and care must be taken to prevent transmission of these pathogens to areas where they do not occur. The most common seed-borne pathogens are tobamoviruses (TMV, ToMV, and PMMV) and bacterial spot [caused by *Xanthomonas campestris* pv. *vesicatoria* (Xcv)]. Tobamoviruses can be detected by surveying plants in the field, using serological tests (leaf symptoms may not be specific for tobamoviruses), grinding up seeds in buffer and performing an ELISA test, or inoculating the seed buffer onto a susceptible local lesion host, such as tobacco. Bacterial spot can be detected by surveying plants in the field for the pathogen, plating seeds on semi-selective media, by polymerase chain reaction (PCR), or by infiltrating seed extract into susceptible pepper leaves. Soaking infected seeds in a 10% (w/v) solution of trisodium phosphate (Na_3PO_4) for 2 hours will eliminate virtually all tobamoviruses and most Xcv. If seeds are only Xcv-infected, they can be soaked in 1.3% (v/v) acetic acid with shaking for 4 hours, rinsed with water three times, and then soaked in 25% Clorox ($4 \times$ dilution) for 5 min and rinsed under running water for 15 min.

Pepper seeds occasionally exhibit dormancy, but this is genotype-dependent (Randle and Honma, 1981; Lakshmanan and Berke, 1998). A dry seed alter-ripening treatment of 21 days at 25°C in the dark has been shown to increase seed germination in lines exhibiting dormancy (Randle and Honma, 1981; Ingham et al., 1993). Pepper seeds may be stored at 25°C in sealed containers for one year with only a 6-12% decrease in percent germination (Thanos, Georghiou, and Passam, 1989). High temperatures (>25°C) and high RH contribute to rapid loss in pepper seed viability. Storage conditions of 15°C and 45% RH can preserve pepper seed germination for at least 10 years (Berke, unpublished data). Pepper seeds have been shown to live up to 50 years under optimum seed storage conditions (5% seed moisture and −18°C in sealed moisture-proof containers). Pepper seeds have a median viability period of < 30 years, comparable to onion and carrot, which is relatively low compared to other vegetables (Roos and Davidson, 1992).

SEED PRODUCTION AND COST

It is difficult to get hybrid seed production statistics because many companies do not publish them. Private seed companies produce hybrid seed worldwide, primarily in America, Chile, China, India, Korea, Peru, Vietnam, and Thailand. Location of hybrid seed production fields depends primarily on the availability of cheap, skilled labor and proper growing conditions (optimum soil type, temperature, and availability of irrigation). Minor amounts of hybrid seed are produced in Taiwan, Israel, Holland, Kenya, and Mexico. Price varies depending on the company, the hybrid, and the country where it is sold. Hybrid seed prices can range from $300-$25,000 kg^{-1}.

REFERENCES

Aleemullah, M., A.M. Haigh and P. Holford. (1996). Anthesis, pollination, gameto-phytic receptivity, and fruit development in chilli pepper (*Capsicum annuum* L.). Proceedings of 13th National Pepper Conference, Naples, Florida. pp. 12-13.

Andrews, J. (1995). Peppers: The Domesticated *Capsicums*-New Edition. Austin, Texas: University of Texas Press.

AVRDC. (1999). AVRDC 1998 Report. Asian Vegetable Research and Development Center, Shanhua, Tainan, Taiwan.

Berke, T.G., and L.M. Engle. (1997). Current status of major *capsicum* germplasm collections worldwide. *Capsicum* and Eggplant Newsletter 16:76-79.

Bettencourt, E. and J. Konopka. (1990). Vegetables-*Abelmoschus, Allium, Amaranthus*, Brassicaceae, *Capsicum*, Cucurbitaceae, *Lycopersicon, Solanum*, and other vegetables. Directory of Germplasm Collections. IBPGR, Rome, pp. 98-129.

Cavero, J., R. Gil Ortega and C. Zaragoza. (1995). Influence of fruit ripeness at the time of seed extraction on pepper (*Capsicum annuum*) seed germination. Scientia Horticulturae 60:345-352.

Cochran, H.L. (1936). Some factors influencing growth and fruit-setting in the pepper (*Capsicum frutescens* L.). Memoirs Cornell University Agricultural Experiment Station 190:3-39.

Cochran, H.L. (1938). A morphological study of flower and seed development in pepper. Journal of Agricultural Research 56:395-417.

Cochran, H.L. and A.H. Dempsey. (1966). Stigma structure and period of receptivity in pimientos (*Capsicum frutescens* L.). Proceedings of American Society of Horticultural Science 88:454-457.

Cotter, D.J. (1980). A review of studies on chile. New Mexico Agricultural Experiment Station Bulletin 673.

Daskalov, S. and J.M. Poulos. (1994). Updated *Capsicum* gene list. *Capsicum* and Eggplant Newsletter 13:15-26.

Demir, I. and R.H. Ellis. (1992). Development of pepper (*Capsicum annuum*) seed quality. Annals of Applied Biology 121:385-399.

Dumas De Vaulx, R. and M. Pitrat. (1977). Interspecific hybridization between *Capsicum annuum* and *Capsicum baccatum*. *In* EUCARPIA Third Meeting on Genetics and Breeding on *Capsicum* and Eggplant, ed. E. Pochard. Montfavet, France, pp. 75-81.

Edwards, R.L. and F.J. Sundstrom. (1987). Afterripening and harvesting effects on Tabasco pepper seed germination performance. HortScience 22:473-475.

Greenleaf, W.H. (1986). Pepper Breeding. *In* Breeding Vegetable Crops, ed. M. Bassett. Westport, Connecticut: AVI Publishing Co, pp. 67-134.

Heiser, C.B., Jr. (1973). Seed to Civilization: The Story of Man's Food. San Francisco, California: W.H. Freeman and Co.

Hirose, T. (1965). Fundamental studies on the breeding of pepper. Technical Bulletin 2. Laboratory of Olericulture, Faculty of Agriculture, Kyoto Prefectural University, Japan, pp. 1-180.

Ingham, B.H., T.C.Y. Hsieh, F.J. Sundstrom and M.A. Cohn. (1993). Volatile compounds released during dry afterripening of Tabasco pepper seeds. Journal of Agricultural and Food Chemistry 41:951-954.

Kanwar, J.S. (1995). Natural cross-pollination and its effect in chilli (*Capsicum annuum*). Indian Journal of Agricultural Science 65:448-450.

Lakshmanan, V. and T.G. Berke. (1998). Lack of primary seed dormancy in pepper (*Capsicum spp.*). *Capsicum* and Eggplant Newsletter 17:72-75.

Lippert, L.F., B.O. Gergh and P.G. Smith. (1965). Gene list for the pepper. Journal of Heredity 56:30-34.

Marfutina, V.P. (1975). Obtaining hybrid seeds of sweet pepper without emasculation of the flowers. Plant Breeding Abstracts 45:822. Abstract No. 10288.

Markus, F. (1969). Correlation between the age and fertilization of red pepper pistil. Acta Agronomica Hungaria 18:155-164.

Maynard, D.N. and G.J. Hochmuth. (1997). Knott's Handbook for Vegetable Growers, 4th ed. New York: Wiley.

Meshram, L.D. and M.N. Narkhede. (1982). Natural male sterile mutant in hot chilli (*C. annuum* L.). Euphytica 31:1003-1005.

Murty, N.S.R. and N. Lakshmi. (1979). Male sterile mutant in *Capsicum annuum* L. Current Science 48:312.

Novak, F.J., J. Betlach and J. Dubovsky. (1971). Cytoplasmic male sterility in sweet pepper (*Capsicum annuum* L.). I. Phenotype and inheritance of male sterile character. Z. Pflanzenzucht 65:129-140.

Odland, M.L. and A.M. Porter. (1941). A study of natural crossing in peppers (*Capsicum frutescens*). Proceedings American Society Horticultural Science 38:585-588.

Pathak, C.S., D.P. Singh and A.A. Deshpande. (1983). Male and female sterility in chilli pepper (*Capsicum annuum* L). Capsicum Newsletter 2:102-103.

Peterson, P.A. (1958). Cytoplasmically inherited male sterility in *Capsicum*. The American Naturalist. 92:111-119.

Poulos, J.M. (1993). *Capsicum* L. *In* Plant Resources of South-East Asia No. 8: Vegetables, eds. J.S. Siemonsma and K. Piluek. Wageningen: Pudoc Scientific Publishers.

Quagliotti, L. (1979). Floral biology of *Capsicum* and *Solanum melongena*. *In* The Biology and Taxonomy of the Solanaceae, eds. J.G. Hawkes, R.N. Lester, and A.D. Skelding. Linnean Society Symposium Series Number 7, New York: Academic Press.

Randle, W.M. and S. Honma. (1981). Dormancy in peppers. Scientia Horticulturae 14:19-25.

Roos, E.E. and D.A. Davidson. (1992). Record longevities of vegetable seeds in storage. HortScience 27:393-396.

Rylski, I. (1984). Pepper (*Capsicum*). In CRC Handbook of Fruit Set and Development. Boca Raton, Florida: CRC Press, pp. 341-353.

Rylski, I. and Spigelman, M. (1982). Effect of different diurnal temperature combinations on fruit set of sweet pepper. Scientia Horticulturae 17:101-106.

Sanchez, V.M., F.J. Sundstrom, G.N. McClure and N.S. Lang. (1993). Fruit maturity, storage, and postharvest maturation treatments affect bell pepper (*Capsicum annuum* L.) seed quality. Scientia Horticulturae 54:191-201.

Shifriss, C. 1997. Male sterility in pepper (*Capsicum annuum* L.). Euphytica 93:83-88.

Shifriss, C. and H. Yehuda. (1977). Segregation for prebifurcation shooting, stem length and leaf number of main stem in two crosses of *Capsicum annuum* L. Euphytica 26:491-495.

Smith, P.B. and C.B. Heiser. (1957). Taxonomy of *Capsicum sinense* Jacq. and the geographic distribution of the cultivated *Capsicum* species. Bulletin of Torrey Botanical Club 84:413-429.

Subramanya, S. (1983). Transfer of genes for increased flower number in pepper. HortScience 18:747-749.

Tanksley, S.D. (1984). High rates of cross-pollination in chilli pepper. HortScience 19:580-582.

Tay, D.C.S. (1991). Extraction of seeds of hot peppers using a modified meat mincer. HortScience 26:1334.

Thanos, C.A., K. Georghiou and H.C. Passam. (1989). Osmoconditioning and aging of pepper seeds during storage. Annals of Botany 63:65-69.

Todorov, J. and G. Csillery. (1990). Natural cross-pollination data from Bulgaria. Capsicum and Eggplant Newsletter 8-9:25.

USDA-ARS. (1999). GRIN database. *http://www.ars-grin.gov/npgs/*

Yu, Il-Woong. (1990). The Inheritance of Male Sterility and Its Utilization for Breeding in Pepper (*Capsicum* spp.). Ph.D. dissertation, Kyung Hee University, 69 pp.

Zijlstra, S., C. Purimahua and P. Lindhout. (1991). Pollen tube growth in interspecific crosses between *Capsicum* species. HortScience 26:585-586.

Hybrid Seed Production in Watermelon

Bill Rhodes
Xingping Zhang

SUMMARY. Controlled pollinations are an essential part of watermelon hybrid seed production, and the use of male sterility will reduce labor required. Profitable seed production must employ good cultural practices and timely harvest. Triploid hybrid seed production varies considerably from diploid hybrid seed production. The generation of tetraploid parents has been enhanced considerably with the use of tissue cultural and chromosome doubling agents other than colchicine. Triploid hybrids, with two chromosome sets from the female parent and one from the pollen parent offers more variety to the consumer as well as more strategies for the employment of past resistance. New triploids can be deployed quickly to meet new pest problems and market demands. *[Article copies available for a fee from The Haworth Document Delivery Service: 1-800-342-9678. E-mail address: getinfo@haworthpressinc.com <Website: http://www.haworthpressinc.com>]*

KEYWORDS. Watermelon breeding, *Citrullus lanatus*, diploid hybrids, triploid hybrids, colchicine, dinitroanilines, controlled pollination, tissue culture

INTRODUCTION

Overview of Watermelon Cultivars. Diploid watermelon cultivars have been enjoyed for a long time. In Egypt the watermelon was

Bill Rhodes is affiliated with the Department of Horticulture, Clemson University, E-142 Poole Agricultural Centre, Box 340375, Clemson, SC 29634 (E-mail: BRhodes@clemson.edu).

Xingping Zhang is affiliated with Novartis, Woodland, CA.

[Haworth co-indexing entry note]: "Hybrid Seed Production in Watermelon." Rhodes, Bill, and Xingping Zhang. Co-published simultaneously in *Journal of New Seeds* (Food Products Press, an imprint of The Haworth Press, Inc.) Vol. 1, No. 3/4, 1999, pp. 69-88; and: *Hybrid Seed Production in Vegetables: Rationale and Methods in Selected Crops* (ed: Amarjit S. Basra) Food Products Press, an imprint of The Haworth Press, Inc., 2000, pp. 69-88. Single or multiple copies of this article are available for a fee from The Haworth Document Delivery Service [1-800-342-9678, 9:00 a.m. - 5:00 p.m. (EST). E-mail address: getinfo@haworthpressinc.com].

known to the pharoahs. Even though watermelon originated in Africa, in the ancient civilization of China, record of the watermelon can be traced back more than a thousand years. However, hybrid watermelon cultivars are a product of this century. Hybrids have replaced inbred (open pollinated, or OP) varieties more slowly in the United States than in Japan and China. Inbreeding depression is not a problem in watermelon, and a number of diploid hybrids show little or no advantage over their parents. Little wonder that the hybrid watermelon did not become an overnight success as hybrid corn did. Hybrids have represented an opportunity for a seed company to control seed production of their cultivars. There are two types of watermelon hybrids: diploids and triploids.

The diploid hybrid is the F_1 progeny from two inbred diploid parents. Because all of the traditional OP varieties were inbred lines, these can be use to synthesize diploid hybrids. The triploid hybrid is the F_1 progeny from a tetraploid maternal parent and a diploid pollen parent. Although triploid watermelon production is small compared with diploid hybrid watermelon production, it represents an increasingly more popular fruit type. The triploid, or so-called "seedless" watermelon is virtually incapable of producing viable gametes because 33 chromosomes in the gametophyte generation assort at random during meiosis, and only a small number of cells receive complete sets of 11 or 22 chromosomes.

Hybrid seed are harvested from the "female" parent, i.e., the line chosen to receive pollen from the pollen parent. *Citrullus lanatus* inbreds are predominantly monoecious, i.e., having both male and female flowers on the same plant (Rosa, 1928). A single dominant gene is responsible for the monoecious condition in watermelon. The monoecious nature of the variety chosen as the "female" is not used, i.e., the pollen from the male flowers of the female parent is either prevented from contaminating the female flowers by hand-pollination, or male sterility is exploited.

In contrast, seed of traditional OP cultivars were produced simply by growing an inbred line in an isolation block and harvesting the seed only from that block. Thus, no special care needed to be taken with OP cultivars to control pollination. Plants produced from OP seed could produce again, in isolation, plants that were true-to-type.

Worldwide, indigenous OP cultivars are more common in less developed countries because any grower can save seed and exchange them. OP cultivars are disappearing because hybrid seed are plentiful and the new hybrids are superior in many respects to the older OP cultivars. Sustainable agriculture would be better served if OP varieties were not lost. The role of germplasm depositories in preserving the old varieties is of utmost importance, and adequate public support for germplasm preservation is essential. Futhermore, the existence of seed savers, seed companies that offer heirloom varieties, and amateur breeders serve the public good.

Value of Hybrids. Traditionally, diploid hybrid seeds have been roughly ten times more expensive and triploid hybrid seed 100 times more expensive than OP seed. The additional cost of production of diploid hybrids can be justified. The very best diploid hybrid cultivars will outperform the very best OP cultivars. We have looked at net profit margins for OP and hybrid diploids in replicated trials in the Southern U. S. and concluded that growing hybrids was more profitable. Kalloo (1993) reported the results of several previous studies that indicated diploid hybrids of watermelon can demonstrate heterosis with respect to yield, number of fruit, size and weight of fruit, fruit uniformity, number of female flowers, earliness, and total soluble solids. Diploid hybrid seed germinate as well as, or better than, OP seed, but because of the expense of hybrid seed, transplanting has become routine to avoid seed loss due to adverse weather. Diploid watermelon seed germinate slowly below 21°C and not at all below 16°C.

In triploids, germination is drastically affected by the presence of a triploid embryo inside a tetraploid seed coat (Kihara, 1951). A triploid seed is actually a triploid embryo inside a tetraploid seed coat and requires an even higher temperature (86°F) and controlled moisture and oxygenation to germinate well. Eighty percent germination has been the norm for triploid watermelon seed before recent innovations in germination. The seed coat is such a formidable problem that it is best scarified before germination and removed from the cotyledons after emergence. However, we have found that simply tumbling the seed with ceramic stones or steel balls improve germination (Rhodes et al., 1997).

When triploid seed germinate, the cotyledons are usually deformed, and the seedling grows more slowly than the diploid. After the cotyledon stage, the triploid seedling grows as vigorously as the diploid. Nevertheless, the special care necessary to start the triploid plant increases the cost of triploid watermelon production. But the triploid hybrid has its superior attributes. Heterosis can be demonstrated for field resistance to Fusarium wilt. Fruit number and total soluble solids are traits exhibiting heterosis (Kihara, 1951). Three other characters not mentioned by Kalloo (1993) but very important are the extremely tough rind of triploids and fruit resistance to watermelon fruit blotch (Garrett, Rhodes and Zhang, 1995; Hopkins and Elmstrom, 1995; Rhodes et al., 1996) and their long storage life. Uniformity, tough rinds, and long storage life reduce shipping costs and increase marketing acceptability.

Likewise, ensuring that only pollen from the desired male parent fertilizes the female flowers of the desired female parent also increases the cost of both diploid and triploid hybrid seed.

ESSENTIALS OF HYBRID SEED PRODUCTION

Overview of Industry. Traditionally, in the United States, and probably elsewhere, enough OP cultivar seed could be produced in small isolation plots to supply the entire country. Indigenous bees were plentiful. Skilled labor was not required. Seed production in the west and northwest parts of the United States was preferred to the growing regions in the South and Southwest where intensive production kept pathogens and pests at higher levels. The production of large quantities of hybrid seed was a paradigm shift. Companies sought and found developing countries for hybrid seed production where skilled labor was cheap and plentiful–China, Chile, India, Thailand and other locations. Unfortunately, the international seed market has encountered problems requiring considerable adjustments: (1) New diseases, such as watermelon fruit blotch; (2) Quality control problems, such as seed purity; and (3) and problems related to location. In China and in other countries where many small farmers grow their own gardens near production fields, to produce hybrid seed, it is necessary to eliminate self pollination on and between "female" parent plants and transfer pollen only from "male" plants. Traditionally, this has been a labor intensive task performed in environmentally suitable countries where labor was abundant and cheap such as China or Chile. A high ratio of male flowers to female flowers–varying between 7:1 and 15:1 depending on field conditions (Feher, 1993)–exists on both parental lines in watermelon. In some *Cucurbita* species, growth regulators can be sprayed on the female line to eliminate the male flowers before bloom.

This approach has not been effective with normal monoecious lines of watermelon. Unlike other cucurbits, applied ethylene appears to suppress rather than promote ovary development of watermelon during flower bud differentiation (Christopher and Loy, 1982). It is necessary to understand the pollination process in more detail to fully appreciate hybrid seed production.

Controlled Pollination. A controlled pollination is a pollination of a known female flower with the pollen of a known male flower, with the exclusion of pollen from all other flowers. Female flowers (flowers with ovaries beneath the flower bud) that will open the next morning must be identified the afternoon before they open. They can be identified by their position on the vine and bud coloration. Prospective female flower buds are located near the apex of a branch of the vine. The petals of the prospective bud have begun to yellow but still have a slight green cast to them. The bud must be firm to the touch. If the bud is not firm and the petals are yellow, the bud has probably opened already and been pollinated, then closed back up. This flower should be removed from the vine. All female flowers that are open pollinated should be removed from the vine during the selection of unopened female flowers to prevent abortion of future females on the same branch. If the female flower bud will open the next morning, the flower must be protected from outside pollination by a cover that will keep out insect pollinators. A glassine bag, with a wire twist tie, or even a paper cover created by rolling a inch-wide strip of paper around the end of a pencil and closing the end is usually sufficient. However, it is imperative to ensure that a given cover will be sufficient to prevent insects from getting into the female flower (Figure 1). This cover must also be easily removed for pollination the next morning and be easily reinserted over the flower after pollination.

Male flowers also need to be located the afternoon before pollination the next morning. Male flowers, like female flowers, begin to yellow before opening the next day and bloom in sequence along the vine. Male flowers that will open in sequence with the female flower are often on the second node below the prospective female flower. However, abnormal environmental conditions can change this sequence. It is rare that the male flower above the prospective female flower opens on the same morning. For a controlled pollination, these flowers would have been covered the evening before they opened. For commercial hybrid seed production, male flower buds are collected in

FIGURE 1. A female flower covered (closed) to prevent insect pollination. Covering is done on the afternoon before a controlled pollination is done the next morning.

late afternoon and kept on moist sand until pollination the next morning. The male flowers collected shed pollen earlier than the flowers in the field if they are kept in a warm place in the light. Alternatively, male flower buds may be collected in the early morning before they open. These procedures save the labor involved in covering individual male flowers. When selecting males for a self pollination, it is necessary to insure that the male used is from the same plant. Therefore, it is best to select a male that is on the same vine as the prospective female. The male flower is more easily closed than the female flower, and care must be taken not to damage the female.

Using an andromonoecious inbred as a female parent of a hybrid should be avoided, if possible. Female flowers of an andromonoecious line have stamens and viable pollen (Figure 2). If an andromonoecious line is used as the female parent, emasculation of anthers before covering the female flower is essential. This extra labor will increase the cost of hybrid seed production considerably.

In a large field, finding the flowers which have been covered or closed for the next morning's pollination is facilitated by some kind of

FIGURE 2. Two andromonoecious flowers, with stamens inserted around the stigma in otherwise normal female flowers that have opened at the same node on the same morning. Male flowers, also with pollen-bearing stamens, may open two nodes below the opening female on the same morning or may be found on another branch of the same plant.

marking system. We have used wire flags or simply stiff fence wire segments with a loop at the top and cut long enough to see above the foliage.

A controlled pollination, cross or self, is made in the following manner. On the morning of the pollination, male flowers are taken to the female flower. The cover is carefully removed from the female flower. In the very early morning, the petals of the female may not be open, and may need to be pried open gently. If the bud is still hard, it may need to be marked for recovering in the same afternoon for pollination the following day. Once the flower is open, the cover of the male flower is removed, and the petals are pulled back to expose the anthers. If the anther does not appear to have shed any pollen, the male flower should be discarded, and a more suitable male used. The surface of the pollen-covered anther is rolled over the surface of the stigma, making sure that the stigma is completely covered with pollen (Figure 3). The female flower is carefully recovered, and tagged.

FIGURE 3. A controlled pollination. The cover (lower right) has been removed from the female flower, and the pollen-laden anthers of the petalless male flowers are being carefully rolled over the entire stigmatic surface of the female flower. Then the cover (lower right) will be replaced, and the identification tag attached to the stem below the female flower.

   Tagging is an extremely important step because it identifies the parents and who did the pollination. The following information, in order, should appear on the tag. The first item on the tag is the plot number of the female flower. The plot number of the female is followed by a crossing symbol, either an X for a cross or an X within a circle for a self pollination. If the pollination is a cross, then the X is followed by the plot number of the male. If the pollination is a self, then no plot number follows the circled X. The next item on the tag is the date of the pollination. The date allows an evaluation of earliness of fruit production. The last bit of information on the tag are the initials of the person who did the pollination. A good pollination, like a good painting, merits the name of the artist! The tag should be placed around the stem just above the female flower (Figure 4). Thus, the tag will remain next to the developing fruit. Sometimes the stem tag is defaced or lost, and it is hard to check until harvest. An added precau-

FIGURE 4. A female flower that has been pollinated and re-covered, with the identification tag attached to the vine above the flower.

tion is to place an identical tag on the wire flag used to mark the site of the female flower.

Pollination should begin between 6:30 and 7 a.m. and is futile after 11 a.m. Most successful pollinations are made by 9 a.m. Covering of flowers should begin in the late afternoon.

When the branches of two plants overlap, tagging mistakes can be made, and the plants are probably too close together to obtain maximal fruit set. Therefore, it is necessary to thin plants to arrive at a uniform stand of one plant per hill, with hills given a more generous spacing than a production field. However, commercial hybrid seed production fields use higher density to increase seed yield. Each hill may have two or more plants per hill at the beginning of the season but should only have one at pollination. Thinning should take place in two phases. First, the hill should be thinned to the two largest plants when the plants have reached the four to five leaf stage. The next thinning should take place when the vines begin to run. At this time the hills should be thinned to one plant. In observational and comparison blocks, hills should be thinned to the three healthiest plants while in the expanded cotyledon stage and two plants at the four or five true-leaf stage.

Use of Male Sterility. Either controlled pollinations or the elimination of pollen-producing flowers can allow the transfer of pollen from only one male parent line to a female parent line. One way to eliminate pollen is to use male sterility. Several male sterile genes have been identified in watermelon. One has particular merit. It is denoted simply as *male sterile* (Figure 5). This recessive gene primarily affects the development of the male flower and has little effect on the rest of the plant. No linked seedling marker for this gene has been identified. This gene eliminates all pollen production and effectively produces an ideal female parent. A 1:1 male fertile:male sterile population can be planted, and the male fertile plants rogued prior to pollination when the male fertile plants with normal plump male buds can be distinguished from male sterile plants with poorly developed male buds (Zhang, Skorupska, and Rhodes, 1994). The use of this gene is further enhanced with a seedling marker (Zhang et al., 1996a, b, c). This system exploits a recessive seedling marker which is not linked with

FIGURE 5. A normal male (left) compared with the male sterile mutant (right). The male sterile anthers are much reduced in size and are pale green instead of yellow.

the male sterile gene. The seedling marker is incorporated into the female line and used to identify inbred seedlings that may show up in the planting of the F_1 hybrid seedlings.

Seed Production. Good production practices are used to produce marketable fruit or seed. Land which has not been previously used for cucurbit production is chosen, soil pH and nutrient levels are determined and adjusted, and adequate water is provided to the crop during its development. Phytophagous insects are controlled during the development of the young, vulnerable seedling and thereafter to prevent the spread of diseases such as bacterial wilt or viruses. *Fusarium* wilt as well as foliar diseases are of considerable concern. Anthracnose and gummy stem blight can be carried in the seed and are assayed in seed tests by some companies. At this writing, perhaps the greatest threat to watermelon seed production is bacterial fruit blotch. Bacterial fruit blotch is caused by *Acidovorax avenae* subsp. *citrulli*. Since the 1960s when it reportedly escaped from a plant introduction in collection at Griffin, Georgia to a breeder's field in Leesburg, it has sporadically caused considerable losses worldwide (Latin and Hopkins, 1995; Hopkins et al., 1997). Until recently it was difficult to detect in a lot of seed, and 10,000-seed growouts are recommended to ensure detection. Now, DNA probes of washings from large seed samples can detect it more easily. The detection of bacterial fruit blotch has been a growth industry generated by the previous growth industry in lawsuits by growers and middle men caught up in the losses associated with this disease. In the seedling stage, in a warm, humid environment, the disease begins as watersoaked spots on the underside of the cotyledon and progresses to large, watersoaked lesions (Figure 6) that become necrotic. If the seedling survives the initial onslaught of the disease, it may develop quite normally thereafter and produce a vigorous plant with a normal load of fruit. As the fruit develops, again under humid conditions, the classic symptom of the disease–a watersoaked lesion rapidly increasing in size–reappears on the upper surface of the developing fruit (Figure 7). Mature fruit are not infected by the disease. As the lesion becomes necrotic, secondary microbes invade the wound and ferment the fruit, leading to gas production and, sometimes, an abrupt cracking of the fruit by the pressure of the gas. The breeder has termed an abrupt cracking of the watermelon fruit "explosive" for a long time, but it took uninitiated members of the press to build a myth of "exploding" fruit to greater proportions. Nevertheless, this disease

FIGURE 6. Watermelon seedling infected with bacterial fruit blotch caused by *Acidovorax avenae* subsp. *citrulli*. Lesions begin as water-soaked spots on the underside of the cotyledon. If conditions are not favorable for rapid growth, the bacterial infection may remain latent until field conditions are favorable for its spread.

has exploded our feelings of security about watermelon seed that facilitated free exchange of germplasm worldwide, and more incredibly, the largely unregulated shipments of large amounts of seed internationally from field to farmer. At this writing, commercial companies are testing 10,000 to 50,000 seed per commercial seed lot for this disease before marketing the seed. This disease can also infect melon.

The average commercial spacing of 2.3 m^2 watermelon should be decreased for seed production. Closer spacing will tend to increase seed yield. High levels of fertility and water are normal for maximal yields of fruit, but frequent watering of watermelon is not necessary nor desired to produce high quality seed. If the soil is not compacted, the roots of the watermelon will penetrate a large volume of soil around the plant and the plant will respond very well to thorough drip irrigations two weeks apart, depending on rainfall. Furrow irrigation is water wasteful and can lead to salt buildup. High quality seed are best produced in latitudes north of fruit production areas where longer days

FIGURE 7. 'Charleston Gray' watermelon infected with bacterial fruit blotch caused by *Acidovorax avenae* subsp. *citrulli*. This seedborne disease is epidemic in rainy seasons or in overhead irrigation. The watersoaked appearance deteriorates as secondary microbes invade the dead rind tissue.

and short, cool nights facilitate the use of photosynthate for the development of the fruit and seeds instead of respiration. Prudent and timely reduction of nitrogen at fruiting will allow timely fruit set and development instead of continued vine growth. Long days, dry conditions and a large day/night temperature differential are particularly important for triploid watermelon seed production. Triploid watermelon seed with more than 90% germination are produced in areas with these environmental conditions.

Currently, very limited production of hybrid watermelon seed employs male sterility. Most of the diploid hybrid seed are produced by hand pollination. Thus, at this moment, an important consideration for choosing a hybrid seed production site is the availability of skilled labor for the 3-4 week pollination period. At least six skilled workers are needed for pollinating one acre of a hybrid watermelon seed production field.

Female and male parents of the hybrid are planted in a ratio of 10 female to 1 male. The male parent is planted 7-10 days earlier than the female parent to ensure adequate pollen at the pollination time. The male parent plants are removed from the field after pollination is complete. Therefore, the male parent is usually planted in outside rows or in the corner of the field.

In commercial hybrid seed production, the pollinated flowers are marked with colored plastic bands or string. A different color is used every 5 days so harvest can be made according to the time of pollination.

Fruit for hybrid seed production are not harvested until over-ripe, usually 7 weeks after pollination. Only marked fruit are harvested for seed. It is highly recommended that a well-trained individual carefully inspect the planting for disease. Discarding diseased fruit, even the entire field if necessary, is the best way to prevent seedborne disease. In addition, 24 hour fermentation of diploid seed before washing almost entirely eliminates the bacterial pathogen for fruit blotch.

The primary method of producing triploid hybrid watermelon seed production has been to employ cheap labor to make controlled pollinations of tetraploid rows with pollen from the diploid male parent. In California, triploid seed are produced by picking off the male flower buds of the tetraploid female row for 3-4 weeks and then marking the fruit set during this time period. If the labor force is well-trained and reliable, and the parents are homozygous, one expects a high yield of pure hybrid seed. By employing genetic marker systems that distinguish hybrid plants from inbred plants before harvest, it is possible to identify and rogue inbred plants before fruit production and thereby produce a uniform crop.

Wall (1960) showed that autotetraploids of very recent origin produce a low percent functional gametes. In addition, the pollen tube of $2x$ pollen grows slower than the pollen tube of $1x$ pollen. Therefore, when tetraploids and diploids are planted together, a high proportion (up to 86%) of the seed harvested from the tetraploid are triploids. The method of mix-planting tetraploids and diploids, allowing open pollination and harvesting seed from open-pollinated tetraploid fruits is still used by some U.S. seed producers.

It is noteworthy that some, if not all, tetraploids are unstable. A small proportion of tetraploids can revert to diploids during reproduction. It is important to eliminate these diploids from the tetraploid population. Seed yield, as well as germination and vigor, is a major consideration for triploid watermelon seed production. Our unpublished experiment showed that fertility of the tetraploid parent is the key factor among factors evaluated for seed yield. Some Chinese researchers suggest the use of a tetraploid hybrid as the female parent to increase triploid seed yield. A major objective in developing a tetraploid is to select for high fertility.

Zhang et al. (1995) demonstrated that the tetraploids induced from the same genotype vary considerably. Tetraploids producing 100 or more seed per fruit are considered adequately fertile. It is possible to find tetraploids producing 200 seed per fruit.

Prompt removal of off-type plants is essential to maintain the uniformity of the cultivar. It is important to be able to distinguish quickly and accurately each cultivar that a company grows and to know the differentiating features of hybrids between these cultivars.

A seed producer is ill-advised to prematurely advertise a cultivar before seed are on hand or to advertise more cultivars than can be efficiently maintained. Cross pollination and off-types are always a threat to the demise of a promising cultivar because growers have been frustrated in their attempts to get the seed or dismayed by the variability in the crop from the seed they plant.

Tetraploid Parent for Triploids. As noted previously, there are two types of commercial watermelon hybrids. *Diploid* hybrids are produced using two diploid inbred lines as parents. *Triploid* hybrids are produced using a tetraploid female parent and a diploid pollen parent.

Tetraploids can be induced by applying aqueous colchicine solution to the growing apex of diploid seedlings or by soaking diploid seeds in colchicine solution prior to germination. Most, if not all, of the tetraploids available have been developed by this method. However, the frequency of tetraploids from such treatments are less than 5% in most cases, and many plants are chimeras. Moreover, fertility of tetraploids developed in this manner is considerably lower than the diploid parents, and these tetraploids may be completely sterile (Stoner and Johnson, 1965). Likewise, spontaneous regeneration of watermelon tetraploids in tissue culture is low (Zhang, Rhodes and Adelberg, 1994; Zhang et al., 1995).

Zhang et al. (1995) reported a protocol for the use of tissue culture in the production and evaluation of tetraploid parents in a 2-year period. The first step is to generate tetraploids by seven days' culture of the interior proximal portion of the cotyledons from fruits harvested 20 to 25 days after pollination on shoot organogenesis medium with

0.05% colchicine. Then the cotyledon tissue is transferred to shoot organogenesis medium without colchicine until shoots are fully developed. After shoot development, the singulated shoots are transferred to MS medium with 5 μM BA for increase or to 5-10 μM IBA for two weeks for root development. The rooted plantlets are acclimatized in the greenhouse where tetraploid plants can be tentatively identified by leaf morphology and flower size. Confirmation of tetraploidy can be obtained by comparing the size of the pollen grains (about 1.44 × larger than diploid pollen, and the number of colpi [4 versus 3, Figure 8]). The confirmed tetraploids are then pollinated with selected diploids to check female fertility by counting the seed number in the mature, hand-pollinated fruit. Axillary buds are collected from the more fertile tetraploids and cultured *in vitro* in the off-season to produce clones of the elite tetraploids. Finally, seed are produced from the cloned tetraploids in an isolation plot, and, simultaneously, the triploids made from the tetraploid are evaluated.

Dinitroaniline herbicides can be used to generate tetraploid plants from diploid plants of many species in tissue culture (Hanson and Andersen, 1996). Dinitroanilines have been used to generate watermelon tetraploids in our lab and by others, but results have not been published. Dinitroanilines are less toxic and induce tetraploidy at concentrations ca. 10^2 lower than needed for colchicine (Hanson and

FIGURE 8. Pollen from tetraploids is 1.44 × larger than diploid pollen and has four pores (colpi) instead of three.

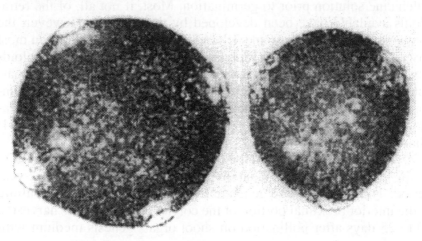

Andersen, 1996). Thus, a considerable potential advantage exists for dinitroanilines when the cheapest, most effective, and least toxic dinitroaniline is determined.

CHARACTERISTICS OF DIPLOID AND TRIPLOID HYBRIDS

Cytogenetic differences exist between diploid and triploid watermelon. The diploid hybrid has two sets of 11 chromosomes that behave normally at meiosis. The triploid watermelon hybrid has three sets of 11 chromosomes that segregate at random. Normal gametes with 11 and 22 chromosomes, respectively, are produced at the rate of $(1/2)^{11}$. Thus, some triploid hybrid fruit generated by diploid pollen in a commercial production field, will have a few viable diploid, triploid and aneuploid seed. The genotype of these offspring will reflect the parents of the triploid as well as the pollenizer parent.

Triploid seed are more difficult to germinate than diploid seed. The temperature during the entire germination stage should be at least 27°C and preferably closer to 30°C. Once the seedling is up and growing the thermostat can be lowered a little to conserve energy. The use of a well-controlled, well-ventilated incubator is recommended for germination. Seeds need constant high temperature and moist, but not wet, conditions to germinate. Air is needed during the germination process. Trays can be covered with polyethylene during the germination process and then removed. Germination rate can be increased by tumbling the seed (Rhodes et al., 1997). The tumbling process is the most significant variable, not unlike the process of scarification. Hopkins et al. (1997) demonstrated that HCl can virtually eliminate bacterial fruit blotch contamination of diploid seed. Perhaps a low concentration of HCl in combination with tumbling may enhance germination as well as protection from bacterial fruit blotch.

The seed coat of the triploid seed is very thick and hard and sometimes sticks to the cotyledons after they emerge from the soil. The cotyledons must be exposed to the light for the seedling to grow. It is very difficult and time-consuming to remove all these seed coats by hand. However, if you spray emerging seedlings with a moderately high pressure nozzle, most of the seed coats will wash off. Maynard (1989) reported that if the pointed end of the seed is planted down at a 45-90 degree angle, the seed coat problem is reduced, but not eliminated. The only mechanical device of which we are aware plants seeds flat, but future models could plant them at an angle.

The flowers of the triploid watermelon have noteworthy characteristics. The triploid plant produces as many male flowers as the diploid, but these male flowers are sterile. Unfortunately, pollinators are still attracted to these flowers and spend more time on these sterile males than they do on the female flowers.

The triploid fruit of the watermelon has a larger blossom-end scar than the diploid fruit, as well as color and shape characteristic of the female parent rather than the diploid pollenizer. If a pollenizer type showing some contrast to the triploid fruit is chosen, it is simple to separate diploids and triploids in the fruit production field. The fertile diploid hybrid does not need a pollenizer.

Hybrid vigor in diploid and triploid hybrids of watermelon has already been noted. Two distinct differences exist between diploid and triploid hybrids. The triploid hybrid has an unequal contribution of genes from the female parent. The tetraploid parent cannot be used as the pollenizer. The tetraploid parent contributes twice as many genes to the triploid than the diploid, and, in the case of anthracnose, for example, the tetraploid parent needs to be the resistant parent because gene dosage plays an important role in resistance to this disease (Henderson and Jenkins, 1977). Other gene dosage effects probably also exist that can be seen in triploids and it is worthwhile to test all the diploid parents as tetraploids.

The triploid watermelon has a tougher rind for shipping, is generally superior in resistance to watermelon fruit blotch, and parents can be chosen to combine disease resistance, flesh color and quality, earliness, yield and even rind color. For example, we have combined an attractive green stripe with a yellow rind to produce a yellow rind with an orange stripe. In short, it is possible to tailor a variety to a particular location and a clientele with a definite fruit type in mind.

Hybrids and Diversity

Since the '70s, production of hybrid seed of watermelon has dramatically increased, driven by the higher value of hybrids in the marketplace. Seed companies and growers have thus far realized a greater return. Unfortunately, the motivation behind hybrid research is more about increased profits than sustainable production. In China the worldwide demand for hybrid seed has fostered the abandonment of seed production of native varieties. The number of seed-saving farmers has diminished. The marker genes, for less labor-intensive hybrid seed production, have become more intensively researched than genes for adaptability and disease resistance. The maximum value of a superior genotype can be realized when it is finally cloned in large numbers, as is the case in Japan, Taiwan and Korea. Mechanized micropropagation systems will generate an almost infinite number of identical clones to be planted where a large number of varieties were used. The obvious result is genetic vulnerability. By offering both OP varieties and hybrid varieties, seed companies may reduce their bottom line but contribute toward maintaining genetic diversity. If a small quantity of some of the old varieties were maintained by each seed company for homeowners, organic farmers and some commercial scale growers, and the unique characteristics of each OP variety were clearly identified, genetic diversity in watermelon could be more easily maintained for crises. In concert with the germplasm units overburdened with the task of characterizing and maintaining too much germplasm with too few funds, private seed companies and public breeders could slow the decline of diversity in this important crop. If genomes can be more fully identified, maintained and deployed during a threat from a pest or pathogen, growers need not endure longterm assaults on their livelihood such as occurred from '89 to the present with watermelon fruit blotch.

REFERENCES

Christopher, D. A., and J. B. Loy. (1982). Influence of foliarly applied growth regulators on sex expression in watermelon. *Journal of American Society for Horticultural Science* 107:401-404.

Feher, T. (1993). Watermelon *Citrullus lanatus* (thunb.) Matsum and Nakai. In: *Genetic Improvement of Vegetable Crops*, G. Kalloo and B. O. Bergh eds. Oxford, Pergamon Press: pp. 295-314.

Garrett, J. T., B. B. Rhodes, and X. P. Zhang. (1995). Triploid watermelons resist fruit blotch organism. *Cucurbit Genetics Cooperative* 18:56-58.

Hansen, N. J. P., and S. B. Andersen. (1996). *In vitro* chromosome doubling potential of colchicine, oryzalin, trifluralin, and APM in *Brassica napus* microspore culture. *Euphytica* 88:159-164.

Henderson, W. R., and S. F. Jenkins, Jr. (1977). Resistance to anthracnose in diploid and polyploid watermelons. *Journal of American Society for Horticultural Science* 102:693-695.

Hopkins, D., R. E. Stall, R. Latin, J. Rushing, W. P. Cook, and A. P. Keinath. (1997).

Research Update: Bacterial Fruit Blotch of Watermelon. *Citrus & Vegetable Magazine*, February, 1997.

Hopkins, D. L., and G. W. Elmstrom. (1995). Comparison of triploid and diploid watermelon cultivars and breeding lines for susceptibility to bacterial fruit blotch. G. Lester & J. Dunlap et al. (eds.), *'Cucurbitaceae 94'*:152-154.

Kalloo, G. 1993. *Vegetable Breeding*. Vol 1. Chap 6. p. 177. Boca Raton, FL: CRC Press, Inc., p.177.

Kihara, H. (1951). Triploid watermelons. *Proceedings of American Society for Horticultural Science* 58:217-230.

Latin, R. X., and D. L. Hopkins. (1995). Bacterial fruit blotch of watermelon: the hypothetical exam question becomes reality. *Plant Disease* 761-765.

Maynard, D. N. (1989). Triploid watermelon seed orientation affects seedcoat adherence on emerged cotyledons. *HortScience* 24(4):603-604.

Rhodes, B. B., C. Huey, A. I. Abramovitch, X. P. Zhang, and T. B. Platt. (1997). Enhancement of triploid watermelon seed germination. *1995-96 Clemson University Vegetable Report*, Clemson, SC, pp. 76-79.

Rhodes, B. B., X. P. Zhang, J. T. Garrett, and C. Fang. (1996). Watermelon fruit blotch infection rates in diploids and triploids. *Cucurbit Genetics Cooperative* 19:70-72.

Rosa, J. T. (1928). The inheritance of flower types in *Cucumis* and *Citrullus*. *Hilgardia* 3:233-250.

Stoner, A. K., and K. W. Johnson. (1965). Overcoming autosterility of autotetraploid watermelons. *Proceedings of American Society for Horticultural Science* 86:621-625.

Wall, J. R. (1960). Use of marker genes in producing triploid watermelons. *Proceedings of American Society for Horticultural Science* 76:577-581.

Zhang, X. P., B. B. Rhodes, and J. W. Adelberg. (1994). Shoot regeneration from immature cotyledons of watermelon. *Cucurbit Genetics Cooperative* 17:111-115.

Zhang, X. P., B. B. Rhodes, H. T. Skorupska, and W. C. Bridges. (1995). Generating tetraploid watermelon using colchicine in vitro. G. Lester & J. Dunlap et al. (eds.), *'Cucurbitaceae 94'*: pp. 134-139.

Zhang, X. P., B. B. Rhodes, W. V. Baird, H. T. Skorupska, and W. C. Bridges. (1996a). Phenotype, inheritance and regulation of expression of a new virescent mutant in watermelon: juvenile albino. *Journal of American Society for Horticultural Science* 121:609-615.

Zhang, X. P., B. B. Rhodes, W. V. Baird, W. C. Bridges, and H. T. Skorupska. (1996b). Development of genic male-sterile lines with juvenile albino seedling marker. *HortScience* 3:426-429.

Zhang, X. P., B. B. Rhodes, W. V. Baird, W. C. Bridges, and H. T. Skorupska. (1996c). Development of genic male-sterile watermelon lines with delayed-green seedling marker. *HortScience* 31:123-126.

Zhang, X. P., H. T. Skorupska, and B. B. Rhodes. (1994). Cytological expression of the male-sterile *ms* mutant in watermelon. *Journal of Heredity*. 85:279-285.

Hybrid Seed Production in Onion

C. S. Pathak

SUMMARY. Onion, *Allium cepa* L., is grown worldwide for its fleshy bulbs which are used as food and medicinal purposes. Based upon global review of the major vegetables, onion ranks second to tomato in area under cultivation. Isolation of male sterility in cv. 'Italian Red' onion led to the development of many hybrid cultivars for various geoecological regions. Although the development of onion hybrid cultivars started in the early 1930s, popularity of onion hybrid varieties is still continuing. In fact, almost two-third of onion varieties in catalogues of major seed companies are listed under hybrid category. This review deals with the floral characteristics, male sterility, production of hybrid onion and method for onion seed production. *[Article copies available for a fee from The Haworth Document Delivery Service: 1-800-342-9678. E-mail address: getinfo@haworthpressinc.com <Website: http://www.haworthpressinc.com>]*

KEYWORD. Alliaceae, *Allium cepa*, male sterility, onion, hybrid seed production, F_1 hybrid seed

INTRODUCTION

Onion, *Allium cepa* L., a member of family Alliaceae (Dahlgren, Clifford and Yeo, 1985) is an important vegetable crop grown worldwide. Its fleshy bulbs are used both in fresh and dehydrated form. The

C. S. Pathak is Manager, Research and Development, Nath Sluis Limited, Nath Road, P.O. Box 318, Aurangabad 431005, India.

[Haworth co-indexing entry note]: "Hybrid Seed Production in Onion." Pathak, C. S. Co-published simultaneously in *Journal of New Seeds* (Food Products Press, an imprint of The Haworth Press, Inc.) Vol. 1, No. 3/4, 1999, pp. 89-108; and: *Hybrid Seed Production in Vegetables: Rationale and Methods in Selected Crops* (ed: Amarjit S. Basra) Food Products Press, an imprint of The Haworth Press, Inc., 2000, pp. 89-108. Single or multiple copies of this article are available for a fee from The Haworth Document Delivery Service [1-800-342-9678, 9:00 a.m. - 5:00 p.m. (EST). E-mail address: getinfo@haworthpressinc.com].

importance of onion as food and medicine is well documented. But the origin of onion still remains a mystery. It has been suggested that it originated in Central Asia (Vavilov, 1951). It is known, however, that cultivation of onion dates back to prehistoric times. References to onion can be found in the Bible, Koran and in the inscriptions of the ancient civilization of Egypt, Rome, Greece, and China. As onion culture spread, cultivars evolved with more diversity in shape, color, flavor, keeping-quality, and with critical adaptations to new climates. The most important adaptive traits involved bulbing response to day-length and high temperature, and bolting response to low temperatures (McCollum, 1976). Onions show the most diversity in the eastern Mediterranean countries, through Turkmania and Tajikstan to Pakistan and India (Astley, Innes, and van der Meer, 1982). A global review of major vegetables show that onion ranks second in area under cultivation after tomato. Approximately 36 million tons of onion are produced worldwide on approximately 2.5 million hectares (FAO, 1996).

Several onion cultivars were developed through breeding for various geoecological regions. However, an important event in the history of onion crop breeding was the isolation of male sterility in cv. 'Italian Red'. This had a great impact on later onion breeding and led to the development of onion hybrid cultivars (Jones and Clarke, 1943).

Onion seed production in general and hybrid seed production in particular, remains a specialized activity. This is due to the crop's sensitivity to photoperiod and temperature for bulb and seed production. In addition to the requirement of two crops (i.e., mother bulb production followed by the seed crop) for a single cycle of seed production.

Development of onion hybrid cultivars started in the early 1930s, and presently most work on the crop is done in the northern USA, Canada, UK, The Netherlands, Germany, Israel, and Japan. Their popularity is increasing in France, Italy, Hungary, Spain, Australia, and New Zealand (van der Meer, 1994). Superiority of onion hybrids over open-pollinated varieties has been reported by several researchers (Fustos, 1986; Sucu, Margea and Neamtu, 1986; Pathak and Gowda, 1994; Panajotovic and Gvozdanovic, 1995; Thornton and Mohan, 1996). Popularity of onion hybrid varieties is increasing and they now constitute approximately 50% of the total short-day onion cultivars listed by seed companies (Currah and Procter, 1990). More recent catalogues of major seed companies list approximately 65% of onion varieties under hybrid category.

FLORAL CHARACTERISTICS

Onion is a highly cross-pollinated crop. Flowering is sensitive to temperature, photoperiod, and stages of bulb development. Cool temperature plays an important role in flower induction. Optimum temperature required for vernalization is 7-12°C (Brewster, 1994). However, it also varies with the cultivar. Cultivars grown in the tropics generally get vernalized even at 15-21°C. Similarly, the duration of vernalization can also vary with cultivar. A duration of 4 to 6 weeks may be optimum for most of the cultivars. Reviews on onion flowering behavior (Brewster, 1987, Rabinowitch, 1990), and pollination (Currah, 1990) are suggested for detailed information.

Stage of bulb development also plays an important role in flowering. Plants at the young juvenile stage do not respond to temperature. A critical stage, which also varies with cultivar, is necessary to initiate response for temperature and flowering. The larger the bulb size, the more easily it can be induced to initiate flowering.

The Flower Stalk (Scape)

Appearance of the scape (flower stalk) signals the initiation of flowering. The scape is a hollow structure which is slightly swollen in the middle and bears the umbel containing about 50 to 2000 individual flowers (the normal range is 200-600). Number of flower stalks produced per plant can vary from 1 to 20. Plants grown from seed produce one flower stalk, however, plants grown from bulbs can produce more flower stalks.

The flower opening in the umbel is irregular and lasts for 2-3 weeks (Figure 1). Individual flowers consist of six pariant lobes, six stamens, and three carpels united in a pistil. Each carpel contains two ovules. The flowers also contain nectaries which help in attracting insects for pollination. The flowers are protandrous, favoring cross pollination. Anther dehiscence occurs 3-4 days prior to the time when the style attains full length and the stigma becomes receptive. The whole process, from anthesis to withering of petals and anthers, takes about 10 days at 18°C and 5 days at 30°C (Brewster, 1994). The highest seed

FIGURE 1. An umbel with open flowers (note the irregular opening of flowers in the umbel).

set was recorded at 35/18°C (day/night) (25°C mean) (Chang and Struckmeyer, 1976).

MALE STERILITY

Hybrid seed production in onion involves the use of male sterility (Figure 2). The genic cytoplasmic male sterility presently used worldwide in onion for commercial exploitation of heterosis was originally derived from cv. 'Italian Red' by Jones and Emsweller (1937). A second source of cytoplasmic male sterility designated as "T" cytoplasm was discovered in onion by Schweisguth (1973). This male sterile line was found to be different than Jones' "S" cytoplasm, because three independent restorer loci were identified for this cytoplasm. The vast majority of onion varieties restore fertility to "T" cytoplasm lines, which makes this "T" cytoplasm more difficult to use.

In addition male sterility has been observed in several other onion populations, mainly in long-day cultivars, such as 'Pukekohe Longkeep-

FIGURE 2. Umbels from fertile (left) and male sterile (right) plants (note, there is no difference in flower morphology except poorly developed anthers in the flowers of male sterile plant).

er', 'Red Wethersfield', 'Scott County Globe', 'Stuttgarter Riesen', and 'Zittauer Glebe'. These have not been genetically characterized, and their relationship to "S" or "T" cytoplasm needs to be established.

The use of a single cytoplasmic male sterile source by the hybrid onion seed industry, derived from 'Italian Red 13-53', has raised concerns regarding genetic vulnerability. There are also reports which indicate the influence of environmental factors on this male sterility, leading to occasional fertile plants. A recently identified highly-stable cytoplasmic male sterile source derived from tropical short-day germplasm might provide genetic variability in the future (Pathak and Gowda, 1994).

PRODUCTION OF ONION HYBRIDS

As onion is a highly cross-pollinated crop, inbred production is essential to obtain heterosis in a hybrid. However, inbreeding depression in onion is so high that inbreeding of more than three generations generally produces very weak inbreds (Jones and Davis, 1944). Thus, 2-3 generations of inbreeding is generally used to develop partially inbred populations.

Cytoplasmic male sterility in onion is controlled by the interaction of the cytoplasm and the nuclear genes. The male-sterile line, used as the female parent, (A line) in hybrid seed production has a genetic constitution of Smsms, having sterile cytoplasm (S) as well as nuclear alleles (msms). Plants having N cytoplasm are always fertile irrespective of their nuclear alleles. However, S cytoplasm in combination with the Ms allele, either in homozygous or heterozygous condition, will produce a fertile plant.

Maintainer lines (B lines) for male sterility are identified in the population and these generally have Nmsms genetic constitution. For this purpose male sterile plants are crossed with several fertile plants originating from different lines. The F_1 hybrids thus produced are evaluated for pollen fertility. The crosses producing sterile pollen are selected and the male parent of this cross is identified for developing maintainer line. Back-crossing is to an "A" line followed to develop an A line with the nuclear genetic makeup of the maintainer, but sterile cytoplasm.

For hybrid seed production, three parental lines (i.e., A, B, and C) are used. The A line used as the female parent, is male sterile, with genetic constitution Smsms, and the seeds collected from this line are used as F_1 seeds. The B line, which acts as a maintainer for the A line is male fertile and has a genetic constitution of Nmsms. The A and B lines are near-isogenic and differ only in their cytoplasms.

The C line is the fertile pollen parent. It is an inbred that is genetically diverse from the A line. It is produced by selfing the plants in the selected lines for one or two generations and then maintained by sib-mating. To identify a suitable F_1 hybrid for commercial exploitation an A line is crossed with several C lines. The best combination giving maximum heterosis for yield and other desirable traits is identified for commercial production. Combining ability studies are used to identify suitable hybrid combinations.

Maintenance of the male sterile parent (A line), the pollen parent (C line), and maintainer line (B line) in pure form is possibly the most important and critical part of any onion hybrid seed production program.

Low seed yield is often a problem in hybrid seed production. Inbreeding depression, which results in reduced plant vigor of inbred parents, is primarily responsible. Inbreeding, however, is necessary to achieve a high level of uniformity and high bulb yield in the F_1 hybrid. During the inbreeding process, the male sterile line can lose its vigor, resulting in low flower production because of inbreeding depression. This is characterized by reduced size and number of umbels, decreased period of receptivity in individual flowers, and ovule abortion (Shasha'a, Nye and Campbell 1973; Campbell, Lotina, and Pollock, 1968 and Ali et al., 1984). Three-way onion hybrids are used to overcome this problem.

Three-Way Hybrids

Three-way onion hybrid seed production involves two stages. First, the male sterile line with a genomic constitution Smsms is crossed with a genetically diverse onion line having a genetic constitution of Nmsms. This produces F_1 hybrid seeds which produce male sterile progeny but have greater seed production capacity than the A line. Second, the F_1 bulbs thus produced are used as the female parent and crossed to another male parent (C line). The seeds collected from the female F_1 parent are used for commercial production as a three-way hybrid.

It is important to maintain bulb uniformity in three-way crosses. To achieve this, parental lines with phenotypic similarity are selected and combined in such a way as to minimize variation in bulb phenotype (Davis, 1966; Ali et al., 1984).

METHODS OF SEED PRODUCTION

Hybrid onion seed for commercial use is produced either using the bulb-to-seed method or by the seed to seed method. The bulb-to-seed method is more commonly used because of its high seed yield, and ease in roguing of unwanted plants before producing seeds in the next generation. The seed-to-seed method requires larger quantities of basic seed, and relatively pure basic seed stock, because roguing is difficult.

Seed Increase of Parental Lines

The initial seed increase of A, B, and C lines must be carried out in isolation, usually in screen cages (Figure 3). The A and B lines are generally planted in the same cage, while the C line is planted separately in another cage. The size of the cage depends on the amount of seed to be increased. Equal numbers of bulbs of the A and B lines are usually planted in separate rows in the cage. At anthesis a net cage cover is put over each group of plants. It is essential to stake and tie up the seed stems on the outside rows of the cages to prevent them from touching the net and becoming contaminated with foreign pollen carried by insects.

Once anthesis starts, each plant is checked very carefully for fertile pollen production (Figure 4). It is essential to rogue out any pollen-fertile plants in the A line and pollen-sterile plants from the B line. This will ensure the purity of seed. Once all of the plants are checked, a beehive is placed inside the cage to help ensure pollination. Seed harvesting is performed very carefully in lines A and B to avoid mechanical mixing. Line C is also maintained in isolation by use of a net cage.

FIGURE 3. Small scale seed production of parental lines in screen cages.

FIGURE 4. Observing the parental line (A line) for off-type fertile plants, which are immediately removed.

For large-scale seed increase, the parental lines are planted in the open field, under proper isolation. The isolation distance between two fertile onion lines should be greater than 3 km. Stock seed of the A line is produced by planting alternate rows of A and B lines. The field size is adjusted based on the seed requirement. Roguing of off-type plants is carried out in the same way as in the cages. Seeds of the C line are increased in a similar manner, taking care to isolate the site. It is advisable to keep bee hives around the seed production plot to insure pollination.

PRODUCTION OF F_1 HYBRID SEED

Climate and Soil Requirements

Identification of a suitable location is the key to commercial seed production. The seed production area should have low humidity, mild, cool temperatures during initial crop growth, followed by increasing temperatures at later stages of growth. Long, rainy periods or heavy dew and fog increase the risk of diseases like downy mildew (*Peronospora destructor*), purple blotch (*Alternaria porri*), and stemphylium leaf blight (*Stemphylium vesicarium*). While the crop is in flower, clear, bright days are necessary to insure high insect activity for pollination. It is equally important to have hot, dry weather during the harvesting, curing, and threshing of the seed. Specific areas are generally used for producing seeds of short-day and long-day onion hybrids. In the USA, Idaho and Oregon are suitable for producing seeds of long-day cultivars, whereas, seeds of short-day cultivars are produced in the Imperial Valley of California (Jones and Mann, 1963). In Europe, major onion seed production areas are located in Italy, southern France, and Spain, because of warm, dry summers which help in the production of disease-free seed crops. In India, the major onion seed production area is in the state of Maharashtra which is also the major onion producing state.

Onion seed crops can be grown on a wide range of soil types, but light sandy soils should be avoided. Heavier soils are preferable because of their higher water retention capacity. The soil should be free from soil-borne diseases such as pink root (*Pyrenochaeta terrestris*), basal rot (*Fusarium oxysporum*), and nematodes (*Ditylenchus* sp.).

Mother Bulb Production

Mother bulb production practices are generally the same as commercial production practices, except in some cases a higher seed rate is used to prevent the production of very large bulbs, which generally do not store well. Special care should be taken to control major diseases affecting the crop. Seedlings of lines A, B, and C are generally planted in the same field but well separated from each other and with proper labeling to avoid mechanical mixing. Bulb size generally plays an important role in seed production. The larger the mother bulb, the higher the seed yield per plant (Chiru and Banita, 1980; Naurai, 1984), but they do not store as well. Smaller size mother bulbs, when planted at higher plant densities in the field, produce increased seed yield per hectare, despite a reduction in per plant yield (Jones and Mann, 1963; Singh, Singh, and Singh, 1977; Currah, 1981). The desired size of mother bulb also depends on the cultivar or inbred parent; however, bulbs of 4 to 6 cm in diameter are preferred (Jones and Mann, 1963; Currah, 1981).

Mother bulbs are harvested when the tops have fallen. Curing the bulbs for approximately one week will increase their storage life. Tops are pruned after the neck is dried. It is very important to label each inbred line properly before storage. Utmost care must be taken during handling and storage to keep the inbred lines separate. Bulbs are stored until they are planted for seed production.

Storage of Mother Bulbs

Mother bulbs are stored under similar conditions as market crops (Figure 5). Ambient temperature storage is preferred among overwintering cultivars, as these are vernalized by cold ambient conditions after the planting of mother bulbs. Several tropical areas generally lack enough cold weather to induce vernalization, and the cultivars grown in these areas do not generally require very cold temperatures for flower induction. These bulbs are stored at ambient high temperatures before planting in the cool season.

FIGURE 5. Onion bulbs stored in ventilated bins for the seed production.

Generally, the ideal storage temperature to prevent sprouting and rotting is 0°-3°C, however, this is not suitable for inducing flowering. Storage conditions of the mother bulbs, especially temperature and duration, generally affect flowering date, number of umbels, and finally seed yield. Three months storage of bulbs at 10°C is sufficient to induce flowering in most cultivars (Peters, 1990). Mother bulbs for seed production should be planted when the temperature is low to avoid the inflorescence suppressing effects of high temperature.

In the tropics, onions are grown mainly as a winter crop, and mother bulbs are harvested in the spring, stored over the summer, and planted in the autumn. No vernalization treatment is required for tropically adapted lines.

Planting of Mother Bulbs

In warm climates, onion bulbs for seed production are usually planted in the autumn, while in cold temperate climates, certain over-wintering cultivars are planted in autumn and others in the spring.

Bulbs are generally planted in rows 50 to 100 cm apart to facilitate cultural operations (Figure 6). In the case of inbred lines, it is advisable to keep higher plant density to increase seed yield, as per-plant yield will be generally low. High plant density, however, has disadvantages–ventilation around the plant is impeded, drying after irrigation is slow, and heavy dew and moisture make the plants more prone to diseases. Ideally, space the plants sufficiently far apart between rows to provide good ventilation, and maintain a high population by planting the bulbs close together within the row.

Carefully selected bulbs of male sterile (A line) and pollen parent (C line) are planted alternately in the field. Sufficient information is available on the required proportion of male and female plants and their planting arrangement in the production field to optimize seed yields (Franklin, 1958; Jones and Mann, 1963; William and Free, 1974; Currah, 1981). In the conventional method, eight rows of male-sterile parent are alternated with two rows of the pollinator parent. A ratio of 12 male-sterile:two pollinator rows has been suggested by Nye, Waller, and Walters, (1971). In contrast, a ratio of nine male-sterile:one pollinator row gave no detectable reduction in seed yield (William and Free, 1974).

FIGURE 6. A typical hybrid onion seed production field. A ratio of 8 (A line):2 (C line) is followed.

Isolation

Commercial F_1 hybrid seeds are produced in isolation during the optimum season for seed production. Onion is an insect-pollinated, outcrossing species and requires isolation from other onion seed production fields, and even from chance bolters from commercial onion fields. The location of onion seed production fields should be planned well in advance, so that adequate isolation is obtained. Complete isolation in a field-grown crop is practically impossible, as pollination is entirely by insects which are able to carry pollen from field to field over long distances. The greater the distance between onion fields, the lesser will be the amount of outcrossing. The isolation distance from other onion fields should be at least two km; however, three km or more isolation distance is preferable to avoid contamination from wild insects. The best isolation distance for onion hybrid seed production is five km.

Fertilizer and Irrigation

Fertilizer requirements of a seed crop are similar to those of a commercial bulb crop. Availability of adequate moisture combined with high soil fertility are most effective in increasing seed yield. Under low rainfall dry land conditions, high levels of fertility do not improve seed yields (Levy et al., 1981). It is recommended that all the required phosphorous be incorporated in the field before planting. Nitrogen and potassium are applied by side dressing or through irrigation. Adequate nitrogen fertilizer is necessary for maximum yield. Improved seed yield is generally obtained by providing optimum quantities of water (Brown, Wright, and Kohl, 1977; Naurai, 1984). To avoid foliar diseases, furrow or drip irrigation is preferred to sprinklers.

Diseases

Downy mildew is one of the most serious diseases in onion seed crops. Purple blotch and stemphylium leaf blight have been reported as the major diseases of seed crops in India (Pandita, 1994). Anthracnose (*Colletotrichum* sp.) (Galli, 1970; Yamamoto and Naito, 1971) and *Botrytis squamosa* are also reported as serious diseases of onion seed crops. The seed stalk can be seriously infected during prolonged periods of leaf wetness. Appropriate field location, field design, spray schedule, and N fertilizer application are necessary to manage diseases in seed crops. Chemical control measures recommended for commercial bulb crops are generally followed for seed crops.

Roguing

Onion, being a highly cross pollinated crop is always at risk of unwanted pollen contamination by insect pollinators, even when utmost care is taken. Roguing is carried out at the bulb production stage as well as at the time of flowering. Bulb color, shape, foliage type, and seed stalk height are some of the parameters used in identifying off-type plants. In hybrid seed production fields, the fertile plants in the female line are immediately rogued out to avoid contamination.

Pollination

To harvest the maximum amount of high quality seed, the flowering period of the A and C lines must be synchronized. If this does not occur naturally, it can often be achieved by adjusting planting dates. Number of days to anthesis of a line is known to be influenced by storage temperature and by planting date of bulbs (Jones and Mann, 1963; Hesse, Vest, and Honma, 1979; Currah, 1981; Brewster, 1982).

Honey bees are the most important among several insect species that pollinate onion flowers. To ensure proper pollination it is essential to place beehives in the seed production field during the flowering period (Figure 7). Generally, 8 to 10 hives per hectare are enough, although 20 or more hives per hectare have been used (Shasha'a, Nye, and Campbell, 1973; Carlson, 1974; McGregor, 1976).

It is recommended not to place all the hives at one time. A few hives should be placed when 50% of the umbels have open flowers. This is mainly to ensure adequate nectar and pollen to forage so that the bees do not move to other fields in search of food. The rest of the hives are introduced periodically during the entire flowering period of the crop.

Insecticide sprays to control thrips or other insects should be managed carefully. Insecticides used to control these pests may be harmful

FIGURE 7. Beehive placed in the onion hybrid seed production field to ensure pollination.

to honeybees. Besides, some chemicals, although not physically detrimental to bees, can leave a residue on the onion plants and make them unattractive to bees (McGregor, 1976).

High potassium concentration in the nectar has been associated with reduced attractiveness of flowers to bees (Waller, Carpenter, and Zeihl, 1972). Similarly, very high sugar concentration (> 50%) in the onion nectar leads to reduced bee activity (Walters, 1972; Currah, 1981). Excessive high temperature conditions can reduce pollen germination, pollen tube growth, and increase the amount of ovule abortion (Chang and Struchmeyer, 1976).

Seed Harvest

Seed maturity takes approximately 30-50 days after anthesis, depending on the cultivar. Seed stalks of A and C lines are harvested separately when the seeds are mature. Care should be taken to avoid mixing seeds between A and C lines. The C line can be harvested before the A line, earlier than the optimal harvest time (when about 25% of the umbels show few open fruits), so that the valuable hybrid seed can be harvested at its optimal time and free from contamination (Peters, 1990).

Frequently, the seed from the C line is discarded. In this case, male plants are destroyed as soon as the pollination is completed. Seeds collected from the female parent (A line) are commercially marketed as F_1 hybrid seeds.

Onion seed may be harvested either by hand or by machine. Seed is hand harvested when about 25% of the umbels show few open fruits and the black seeds are visible. The umbels are cut with about 15 cm of scape attached. Mechanized harvesting is recommended when ripe seeds are visible in 1-3% of the umbels (Globerson, Sharir, and Elias, 1981).

After harvest, the umbels can be dried in several ways. Sun drying on canvas or plastic sheets is commonly practiced when the weather is clear and sunny. The umbels are spread loosely in a layer about 20 cm thick. To achieve uniform drying and to avoid rotting, the umbels are turned regularly. Umbels can also be dried on racks in sheds, or in bins with forced warm air. To avoid damage, the temperature of the warm air should not exceed 32°C until the seed moisture content is less than 18%; 38°C until less than 10%, and 43°C when below 10% (Brewster, 1994).

Seed is ready to thresh when the capsules and small seed stems are brittle and break readily when rolled in the palm of the hand. Most threshing is done with a combine. Caution must be taken not to injure the seed. The seed is threshed and collected in the bags. Small quantities of seed may be threshed by rolling, flailing, stamping, or rubbing out between boards faced with finely corrugated rubber matting.

From the thresher, the seed is cleaned by fan mills and gravitation tables. Seed cleaning can also be done by immersing the seed in water and floating off the non-seed residues. In this method, the seed must be immediately dried to less than 12% moisture content. Hand cleaning using sieves and wind is also carried out, usually for very small quantities.

Seed yield generally varies and will depend on the vigor of the female parent. Hybrid seed yields of about 500-800 kg ha^{-1} can be obtained.

Seed Storage

Viability of onion seed during storage depends on two factors: (1) moisture content of the seed, and (2) seed storage temperature. At high temperature and humidity, viability of onion seeds is lost within one year. If the moisture content of the seed is reduced to < 6.3% or lower and the seeds are sealed into moisture-proof cans or foil packets, the viability can be maintained for at least three years even in warm climates. Low temperatures likewise increase seed longevity. Seed longevity can be increased by keeping seeds very dry and/or very cold. Low moisture content is easier to achieve and generally used in commercial practice.

REFERENCES

Ali, M., B.D. Dowker, L. Currah, and P.M. Mumford. (1984). Floral biology and pollen viability of parental lines of onion *Allium cepa* hybrids. *American Applied Biology* 104: 167.

Astley, D., N.L. Innes, and Q.P. van der Meer. (1982). Genetic resources of *Allium* species–a global report. IBPGR Secretariat, Rome, p.7.

Brewster, J.L. (1982). Flowering and seed production in over wintered cultivars of bulb onions. I. Effect of different temperatures and day lengths. *Journal of Horticultural Science* 57: 93.

Brewster, J.L. (1987). Vernalization in the onion–a quantitative approach. In *The Manipulation of Flowering*, ed. J. G. Atherton. London: Butterworth. pp. 171-183.

Brewster, J.L. (1994). *Onion and Other Vegetable Alliums*. Wallington, UK: CAB International. pp. 142.

Brown, M.J., J.L. Wright, and R.A. Kohl. (1977). Onion seed yield and quality as effected by irrigation management. *Agronomy Journal* 69: 369.

Campbell, W.F., S.D. Cotner, and B.M. Pollock. (1968). Preliminary analysis of onion seed (*Allium cepa* L.) production problem 1966 growing season. *Hort Science* 3: 40.

Carlson, E.C. (1974). Onion varieties, honey bee variations, and seed yield. *California Agriculture* 28: 16.

Chang, W.N., and B.E. Struckmeyer. (1976). Influence of temperature on seed development of *Allium cepa* L. *Journal of the American Society for Horticultural Science* 101: 296-198.

Chiru, C., and I. Banita. (1980). Experience of the Buzau Vegetable Research and Production Station Romania in onion seed production. *Productia Vegatala Horticulture* 29: 9.

Currah, L. (1981). Onion flowering and seed production. *Scientific Horticulture* 32, 26-46.

Currah, L. (1990). Pollination biology. In *Onion and Allied Crops* Vol. 1, eds. H.D. Rabinowitch, and J.L. Brewster. Boca Raton, Florida: CRP Press, pp. 135-149.

Currah, L., and F.J. Proctor. (1990). *Onions in Tropical Regions*. Bulletin 35. Chatham, UK: Natural Resources Institute.

Dahlgren, R.M.T., H.T. Clifford, and P.F. Yeo. (1985) *The Families of the Monocotyledons*. Berlin: Springer-Verlag. pp. 193

Davis, E.W. (1966). An improved method of production of hybrid onion. *Journal of Heredity* 57: 55.

Food and Agriculture Organization. (1996). *FAO Quarterly Bulletin of Statistics*. 9(3/4): 86-87.

Franklin, D.F. (1958). Effect on hybrid seed production of using different ratio of male sterile and pollen rows. *Proceedings of the American Society for Horticultural Science* 71: 435.

Fustos, Z.S. (1986). Increasing earliness in onions by developing F_1 hybrids. *Zoldsegtermoszteis-Kut-Intez-Bull Kecskemet: Mezogazdasgi Konyu Kiado, Vallalat* 19: 19-24.

Galli, J. (1970). Diseases and fungicides in the production of onion seeds. *Pesqui Agropecu. Bras.* 5: 227.

Globerson, D., A. Sharir, and R. Elias. (1981). The nature of flowering and seed maturation in onions as a basis for mechanical harvesting of the seeds. *Acta Horticulturae* 11: 99.

Hanelt, P. (1990). Taxonomy, evolution, and history. In *Onion and Allied Crops*. Vol. 1, eds. H.D. Rabinowitch and J.L. Brewster. Boca Raton, Florida: CRC Press. pp. 1-26.

Hesse, P.S., G. Vest, and S. Honma. (1979). Effect of 4 storage temperatures on seed yield components of 3 onion inbreds *Allium cepa*. *Scientific Horticulture* 11: 207.

Jones, H.A., and A.E. Clarke. (1943). Inheritance of male sterility in onion and the production of hybrid seed. *Proceedings of the American Society for Horticultural Science* 43: 189-194.

Jones, H.A., and G.M. Davis. (1944). Inbreeding and heterosis and their relation to the development of new varieties in onions. *Technical Bulletin of the U.S. Department of Agriculture* 874: 28.

Jones, H.A. and S.L. Emsweller. (1937). A male sterile onion. *Proceedings of the American Society for Horticultural Science* 34: 582-585.

Jones, H.A., and L.D. Mann. (1963). *Onions and their Allies.* New York: Interscience Publishers.

Levy, D., Z. Ben-Herut, N. Albasel, F. Kaisi, and I. Manasra. (1981). Growing onion seeds in a arid region drought tolerance and the effect of bulb weight, spacing and fertilization. *Scientific Horticulture* 14: 1-7.

McCollum, G.D. (1976). Onion and allies, Allium (Liliaceae). In *Evolution of Crop Plants*, ed. N.W. Simmonds, Longman Press, pp. 186-190.

McGregor, S.E. (1976). Insect pollination of cultivated crop plants. *Agricultural Handbook.* pp. 496.

Naurai, A.H. (1984). View of research on onion (*Allium cepa* L.) seed production in Sudan. *Acta Horticulturae* 143: 99.

Nye, W.P., G.D. Waller, and N.D. Walters. (1971). Factors affecting pollination of onions in Idaho during 1969. *Journal of the American Society for Horticultural Science* 96: 330.

Panajotovic. J., and J. Gvozdanovic-varga. (1995). Investigation on combining ability of inbred lines and expression of heterosis for the average size of onion bulbs. *Genetika (Yugoslavia)* 27: 141-150.

Pandita, M.L. (1994). Status of allium production and research in India. *Acta Horticulturae* 58: 79-86.

Pathak, C.S., and R.V. Gowda. (1994). Breeding for the development of onion hybrids in India: Problems and prospects. *Acta Horticulturae* 358: 239-242.

Peters, R. (1990). Seed production in onions and some other *Allium* species. In *Onion and Allied Crops.* Vol. 1, eds. H.D. Rabinowitch and J.L. Brewster. Boca Raton, Florida: CRC Press, pp. 161-176.

Rabinowitch, H.D. (1990). Physiology of flowering. In *Onions and Allied Crops.* Vol. 1, eds. H.D. Rabinowitch, and J.L. Brewster. Boca Raton, Florida: CRC Press, pp. 151-159.

Schweisguth, B. (1973). Study of a new type of male sterility in onion, *Allium cepa* L. *Ann. Amelior. Plants.* 23: 221.

Shasha'a, N.S., W.P. Nye, and W.T. Campbell. (1973). Path coefficient analysis of correlation between honey bee activity and seed yield in *Allium cepa* L. *Journal of the American Society for Horticultural Science* 98: 341.

Singh, V., V. Singh, and I.J. Singh. (1977). Effect of type of bulb distance and date of planting on the performance of seed crop of onion (*Allium cepa*). *Balwant Vidyapeeth Journal of Agricultural Science Research* 16: 32.

Sucu, Z., R. Margea, and G. Neamtu. (1986). Results concerning the behavior of some promising onion hybrids. *Analele-Institutului-de-carcefari-pentru-Legumicultura-Si Flori-cultura* 8: 143-149.

Thornton, M.K., and S.K. Mohan. (1996). Response of sweet Spanish onion cultivars and numbered hybrids to basal rot and pink root. *Plant Disease* 80: 660-663.

van der Meer, Q.P. (1994). Onion hybrids: Evaluation, prospects, limitations, and methods. *Acta Horticulturae* 358: 243-248.

Vavilov, N.I. (1951). The origin, variation, and immunity and breeding of cultivated plants. *Chron. Bot.* 13: 1-6

Waller, G.D., E.W. Carpenter, and O.A. Zeihl. (1972). Potassium in onion nectar and its probable effect on attractiveness of onion flowers to honeybees. *Journal of the American Society for Horticultural Science* 97: 535.

Walters, N.D. (1972). Honey bee activity in blooming onion fields in Idaho. *American Bee Journal* 112: 218.

William, I.H., and J.B. Free. (1974). The pollination of onion *Allium cepa* to produce hybrid seed. *Journal of Applied Ecology* 11: 409.

Yamamoto, H. and H. Natio. (1971). The perfect stage of an anthracnose of onion seed production. *Kogawa Daigaku Noga Kubu, Gakuzyutu Hokoku.* 23: 5.

Hybrid Seed Production in Cabbage

Fang Zhiyuan
Xiaowu Wang
Qu Dongyu
Liu Guangshu

SUMMARY. Cabbage (*Brassica oleracea* L. *var capitata* L.), an important vegetable crop belonging to the family Cruciferae, is grown in many countries espeically in Europe, North America, and Asia. As per botanical characteristics, it can be classified as common cabbage, heading cabbage, and savoy. It is a cross pollinated crop showing strong heterosis in F_1 hybrid progeny. Due to high yield, strong disease resistance, wide adaptability, good quality and uniform economic characteristics, hybrid cabbage have been widely used in many countries. The present review deals with vegetative and reproductive biology of the crop and the recent approaches for hybrid seed production in cabbage. *[Article copies available for a fee from The Haworth Document Delivery Service: 1-800-342-9678. E-mail address: getinfo@haworthpressinc.com <Website: http://www.haworthpressinc.com>]*

KEYWORDS. Cabbage, Cruciferae, growth and development, hybrid seed production, breeding and male sterility

Cabbage has wide adaptability, high disease and stress resistance, high yield potential and strong transporting tolerance. It is cultivated

Fang Zhiyuan, Xiaowu Wang, Qu Dongyu, and Liu Guangshu are affiliated with the Institute of Vegetables and Flowers, Chinese Academy of Agricultural Sciences, 30 Baishiqiao Road, Beijing 100081, China (E-mail: ivfcaas@public3.bta.net.cn).

[Haworth co-indexing entry note]: "Hybrid Seed Production in Cabbage." Zhiyuan, Fang et al. Co-published simultaneously in *Journal of New Seeds* (Food Products Press, an imprint of The Haworth Press, Inc.) Vol. 1, No. 3/4, 1999, pp. 109-129; and: *Hybrid Seed Production in Vegetables: Rationale and Methods in Selected Crops* (ed: Amarjit S. Basra) Food Products Press, an imprint of The Haworth Press, Inc., 2000, pp. 109-129. Single or multiple copies of this article are available for a fee from The Haworth Document Delivery Service [1-800-342-9678, 9:00 a.m. - 5:00 p.m. (EST). E-mail address: getinfo@haworthpressinc. com].

in many countries, especially those in Europe, North America and Asia. Head cabbage cultivars descended from wild non-heading *Brassica* and originated in the eastern Mediterranean and Asia Minor. Some types of non-heading *Brassica oleracea* are widely cultivated in European countries since 9th century. After artificial selection, the heading types were created in Europe in the 13th century. In the 16th century head cabbage was introduced to North America and China, and to Japan in 18th century (Suteki, 1984; Fang, Sun, and Liu, 1991; Fang, 1993).

Cabbage can be classified as common cabbage, heading cabbage and savoy according to its botanical characteristics. The widely cultivated one is the common cabbage that falls into three different head types, which are pointed head, round head and drum head based on head shape. Point head cultivars are mostly early maturing, have strong immature bolting tolerance resistance and chilling, but are susceptible to disease and heat stress. Round head type has round or nearly round head with characteristics of early or mid-early maturing, hard and crisp foliage head of good quality. But this is prone to immature bolting than point head type if it is sown too early or improperly managed in early spring. It is also susceptible to diseases and heat stress. Most of the drum headed varieties are medium or late maturing, resistant to immature bolting, diseases and heat. When the varieties are cultivated in spring they hardly show any immature bolting.

Cabbage is a cross pollinated crop with strong heterosis in F_1 hybrid progeny. Cabbage hybrids have been used widely in the world because of their high yield, strong disease resistance, wide adaptability, good quality and uniform economic characteristics.

GROWTH AND DEVELOPMENT CHARACTERISTICS RELATED TO SEED PRODUCTION

Phasic Development Characteristics of Cabbage

Cabbage is a biennial crop. The vegetative development occurs during autumn and vernalization occurs at low temperature in winter. After vernalization, bolting occurs followed by flowering in the spring. Cabbage can only be vernalized at the seedling stage. The germinating seeds of cabbage do not respond to vernalization at low temperature. There are two conditions for vernalization. The first is that the cabbage plant must attain a certain size after finishing the juvenile stage, and then exposed to low temperature within a certain range, for a specific time. Generally, the seedlings with more than 0.6 cm of stem diameter and seven real leaves are vernalized at 0 to 12°C for 50-90 days. If the temperature exceeds the range, it shows a weak vernalization response. The temperature sensitivity for vernalization also varies according to the plant size, a larger plant being more sensitive than a smaller one and needs a shorter duration of low temperature exposure.

The vernalization response varies greatly among different cabbage cultivars and types. To be vernalized, different varieties need not only different duration of low temperature exposure but also different plant size. Most point-headed and drum-headed varieties are hard to be vernalized and have less chances of immature-bolting in spring cultivation. The converse is true in case of round-headed varieties.

The vernalization of varieties with strong bolting resistance needs a stem diameter larger than 1.5 cm, leaf number more than 16, and a 90-day duration of low temperature exposure. So they can live through winter and immature-bolting rarely occurs. For weak bolting resistant varieties, the vernalization needs only a stem diameter more than 0.6 cm, leaf number more than 7, and 50-60 day duration of low temperature exposure. For these varieties, measures such as late sowing, raising seedlings with a controlled size are needed to prevent immature-bolting.

As to the effect of day length on the bolting of cabbage, long days promote bolting, but the photoperiodic response varies greatly in different ecotypes. Generally, point-headed and drum-headed varieties are less sensitive to day length than round-headed varieties. Stock plants of point-headed and drum-headed varieties buried underground or stored in a cave during the winter can bolt normally in the following spring. Round-headed varieties have a strict need for day length. A great number of plants overwintered without light cannot bolt normally.

High temperatures over 30°C have a vegetative reversion or vernalization reversion effect on cabbage bolting (Fang, 1991). When a vernalized plants planted in an environment with a temperature over 30°C, the buds will stop flowering and the plant reverts to vegetative growth and grows new leaves on the flower stalk. This is often observed in stock plants in greenhouse, especially for round-headed cabbage.

Morphology of Stock Plants and the Branching Habits

The plant height of stock plants varies according to varieties and agronomic management. Early point-headed varieties have a height of 1.0 to 1.2 m, and the round-headed or drum-headed from 1.3 to 1.8 m. Plants growing in protected field over the winter can grow up to one third higher than those transplanted in the field in the next spring.

The first flower branches are generated in the leaf axil of the main flower stalk and the second in the leaf axil of the first ones. Plants growing under sufficient nutrient and good management conditions can generate even the third and fourth branches. Because of their differences in branching habit, the stock plants of different ecotypes vary greatly in plant shape. Round-headed variety has strong main stalk and a few weak branches. In the beginning of bolting, there is only a main stalk. The first branches generate slowly. Compared to the round-headed variety, the point or drum-headed variety has relatively weaker main stalk, but a larger number of strong branches that include even the third ones. In the same variety, stock plants planted before winter generate more branches and are stronger than those planted in the coming spring. When a plant is in poor seed setting or after artificial branch cutting, point or drum-headed types will generate new branches quickly, but this can rarely occur for many round-headed varieties. These characteristics can be applied in hybrid seed production to adjust the flowering time.

Flowering

A healthy cabbage stock plant can generate 800-2000 flowers depending on the variety and agronomic management. A head planted before winter can generate more than 1000 flowers with good management, but only 300-400 flowers when planted in the next spring combined with poor management. The flowering occurs on the main flower stalk first, followed by the first order of branches from top to the bottom and then the second, third and fourth branches. On one flower branch, flowers open orderly from the bottom to the top. No mater how strong its main flower stalk is, the most flowers appear on the first order of branches and less on the second and main stalk.

Flower time varies greatly in different varieties. Generally, varieties with point or drum head, flower 7-15 days earlier than those with round head under similar agronomic management conditions, flowering lasts for 30-50 days in a given variety. Plants or varieties with strong vigor have a relatively longer flowering period than their weaker counterparts. There are 2-5 buds opening each day on one in florescence, 4-5 in clear and warm days, and 2-3 in cloudy or rainy days. A flower opens for 3 days before withering in nature and 4-5 days under humid conditions such as in the greenhouse or an isolating bag.

Biology of Fertilization

Cabbage has a complete flower with calyx, corolla, androecium and gynoecium. Each calyx has four sepals, and the corolla is formed with 4 yellow or golden yellow petals arranged in a cross. Two short and four long stamens are found inside the petals with each having an anther on the top. Mature pollen grains are released by anther dehiscence. At the bottom of the stamens, there are 4 nectar glands. The gynoecium including stigma, style and ovary grows in the middle of a flower. An ovary has two carpels, each containing 20-30 ovules at the two sides. The androecium and gynoecium have the same length at the beginning of flower opening.

Cabbage is a typical, cross-pollinated plant with strong self-incompatibility (Nasrallah and Wallace, 1968). Natural pollination is mediated mainly by insects, especially bees. The percentage of hybrid seeds can be as high as 70% when two varieties are planted together and pollinated naturally. Commonly, 30-40% plants in a variety have a self-incompatibility index less than 1 in artificial selfing test at flowering stage (Fang, Sun, and Liu, 1983). Self-incompatibility provided the basis for the hybrid seed production in cabbage.

Stigma receptivity and pollen viability are highest on the first day of flower opening. A stigma can be fertilized 6 days before or 2-3 days after flower opens. Pollen are viable 2 days before or 1 day after flower opening, but the viability can be maintained for more than 7 days if pollen grains are stored in a desiccated state at room temperature and even longer at temperature below 0°C.

It takes about 36-48 h from pollination to fertilization at the optimum temperature of 15-20°C. When the pollen grains land on the stigma, it takes 2-4 h for the pollen tube to begin elongation; 6-8 h later the pollen tube penetrates the stigma tissue and fertilization accomplished 36-48 h thereafter. Pollen germination slows down below 10°C and the normal fertilization is affected at temperatures higher than 30°C.

Characteristics of Seed-Setting

Seed development of cabbage varies among the different heading types and varieties. The maturing time is longer for round-headed type than for point or drum-headed ones and is faster at high temperature compared with low temperature. There are 600-1500 pods on each plant, depending on management practices. Heads planted before winter have much more available pods than those planted in the next spring. The available pods are concentrated mainly on the first order of branches and then on the second order and main branches.

There are approximately 20 seeds in each pod. The seed number is less in pods on the top, but more in pods on the middle and bottom. The color is reddish brown or dark brown for mature seeds. The weight is 3.3-4.5 g per 1000 seeds. A healthy plant can yield 40-50 g seeds. A dry and cool environment is favorable for seed storage. Fully matured seeds can be stored for 2-3 years under favorable room conditions, but only 1-2 years in humid conditions. If the seeds are stored dry, high germination rate can be ensured for 8-10 years.

THE CHOICE OF SEED PRODUCTION AREA

Since cabbage is a typical biennial plant that vernalizes in winter and flowers in the next spring, two main climatic factors should be taken into consideration while selecting the area for cabbage seed production. The most important one is the temperature in winter that is critical for vernalization. The other one is the precipitation in the season of flowering, seed maturing and harvesting.

The vernalization should occur for a period of more than 60 days at a low temperature. On the other hand, since the temperatures below $-10°C$ are generally injurious to plants, the safe temperatures of stock plant overwintering should be higher than this limit.

In tropical or subtropical areas, the temperature in winter is too high for cabbage to pass through the vernalization stage, and can cause irreversible injury to the overwintering stock plants. So these regions are not suitable for cabbage seed production. Most areas of commercial cabbage seed production are in the Temperate Zone where it is easy to find favorable temperature conditions for vernalization and safe overwintering of cabbage stock plants.

Precipitation mainly affects seed setting. Rain at flowering stage limits the activity of pollinating insects, causes the dropping of flowers, and consequently reduces the seed yield. Excessive rainfall during seed maturity and harvesting may cause many kinds of diseases, seed rot, and pre-harvest sprouting in pods, causing deterioration of the seed quality.

APPROACHES FOR HYBRID SEED PRODUCTION

It is very difficult to produce F_1 hybrid seed in cabbage by hand emasculation and pollination because of its small floral organs. Therefore, the application of scientific, rational and operative methods to produce hybrid seed is the key issue for utilization of heterosis in cabbage. Theoretically, hybrid seed production in cabbage can be achieved through the following approaches.

Intervarietal Crossing or Crossing Between Inbred Lines

Two suitable varieties or normal inbred lines are selected as parents to produce F_1 hybrids. Then the parental lines are planted together at a ratio of 1:1 in the same isolated plot for open pollination. The seeds harvested are F_1 hybrid seeds. However, the hybrid purity can reach only 70%, which has restricted the use of this method in hybrid seed production.

Application of Marker Characteristics at the Seedling Stage

Hybrid seedlings may be distinguished from their parents by differences on morphological characters of the parents and F_1 hybrids, but typical and identical marker characters have not been found in cabbage so far. Even if some characters are identified they can not be used as markers because of their genetic complexity.

Breeding and Utilization of Male Sterile Lines

F_1 hybrid seeds produced can reach 100% purity at lower cost by using male sterile lines as female parents. The first type recessive nuclear male sterility was reported by Cole (1959), Nieuwhof (1961),

Sampson (1966), Borchers (1966). However, no marker genes linked to male sterility were found. Hence it is difficult to use this kind of male sterility. The second type of male sterility controlled by *nigra* nuclear-cytoplasm interaction was reported by Pearson (1972). The defects of this type of materials are that the petals can not open fully and the nectar gland is less developed, and does not attract insect pollinators. The third type, Ogura male sterility, was introduced from radish cytoplasm (Bannerrot, Loulidard, and Tempe, 1974). Its major problem is chlorosis at low temperature and suppressed nectar gland development (Len, 1988). In recent years, the protoplast fusion has been used to improve male sterile lines with radish cytoplasm by scientists in France and USA (Pelletier et al., 1983; Earle et al., 1994). While the chlorosis problem at seedling stage has been solved, the improved materials showed reduced nectar production, inability to attract insects and low seed setting by open pollination. So far it has not been utilized in commercial seed production. The fourth type is the dominant male sterile gene in cabbage. The results from trials of several years verified that male sterility is controlled by a pair of dominant nuclear genes. This male sterile material has been used to develop homozygous dominant male sterile lines. Several superior hybrids from male sterile lines have been developed and will be used for commercial production in the next few years.

Breeding and Utilization of Self-Incompatible Lines

It is quite easy to breed a self-incompatible line in cabbage. Through continuous self-pollination and oriental selection, an incompatible line with very low seed-set or even no seed-set when pollinating between plants within the same line, can be developed. Furthermore, when two self-incompatible lines are used as parents to produce hybrids, the reciprocal crossed seeds can be harvested as hybrids. In 1950, Nagao-ka No.1, which was the first cabbage hybrid in the world, was developed through self-incompatible lines in Japan. From 1960s to 1970s, self-incompatible lines were used extensively in the production of hybrid seed in various countries in Europe, America and Asia.

The superior self-incompatible lines for hybrid seed production should possess the following characters:

Stable Self-Incompatibility: A self-incompatible line requires that each flower set not more than one seed by flower pollination between plants within the same inbred line. Only two self-incompatible lines with stable self-incompatibility are used as parents and for obtaining pure hybrid seed.

High Seed-Set of Self-Pollination at Bud Stage: Self-incompatible lines can not set seeds by selfing at flowering stage, but can do so at bud stage (Pearson, 1931). Therefore, self-incompatible lines can be propagated with self-pollination at bud stage. The better the seed-set at bud pollination is for propagation of self-incompatible lines, the lower the cost of seed production of parental lines. At least 5-10 seeds per flower should be set by selfing at bud pollination.

Favorable and Uniform Economic Characters: Less growing vigor degeneration from selfing can ensure that F_1 hybrids possess superior and uniform economic characters.

Desirable Combining Ability: This can ensure that F_1 hybrids have stronger heterosis.

Until now, almost all cabbage hybrid seeds have been produced with self-incompatible lines all over the world. Thus this review focuses on hybrid seed production using self-incompatible lines. This includes propagation of basic seed for parents and using F_1 seed production as practiced in China.

PROPAGATION OF BASIC SEEDS FOR PARENTS

Nursery of Stock Plants Before Winter

Stock plants can only bolt and flower after vernalization at low temperature, therefore it has to be grown in the autumn of previous year. The sowing time is dependent on maturity of parents. Normally late sowing is recommended because parents have weak stress resistance after continuous selfing. Maintenance of small head has the advantages of reducing disease damage at seedling stage by later sowing. The loose head is formed with resistance to diseases and suitable for overwinter storage. The stock plants grow in the coming year and produce higher yield of hybrid seeds. However, the sowing date can not be too late otherwise the head is so small that proper vernalization can not be achieved. Flowering and seed setting are also affected. Generally, seeds of self-incompatible lines with medium and medium late maturity can be sown for raising seedlings in the mid-July or directly sown in the early August. Seeds of self-incompatible lines with early and medium-early maturity are sown for raising seedlings in early August. Seed beds ought to be selected in fertile soil at higher terrain. Before emergence, seed bed should be covered with shading net to protect the seedlings against storm and sun exposure. Seedlings with 3-4 real leaves require to be transplanted at a plant spacing of 10 cm and finally transplanted into the open fields when 7-8 true leaves have emerged. If complete solid head is used to produce seeds, the plant and row spacing for mid-late maturity lines is 43×50 cm^2 and 33×40 cm^2 for early maturing ones. In case of small head, the plant and row spacing for mid-late maturing lines is 36×43 cm^2 and 27×33 cm^2 for early-maturing ones. Cabbage worms and aphids should be controlled by spraying chemicals in addition to water and fertilizer management during whole growing season.

Stock Plant Selection for Seed Production

Generally, stock plant selection is conducted at the seedling stage, head formation and bolting.

Selection at Seedling Stage: The seedlings are selected as stock plants with no disease infection, vigorous growth and characteristics of inbred lines such as, leaf shape, leaf color, leaf margin, leaf surface wax and leaf stalk, etc.

Selection of Complete Solid Head and Small Head Before Winter: The healthy plants with uniform outer leaves and head are selected as stock plants.

Selection in the Bolting and Flowering Stage: The stock plants are selected according to the branching habit, color of flower stalk and leaflets on the stem, bud shape and bud color at the flowering stage. Off types are eliminated.

Stock Plant Storage in Winter

In temperate zone, stock plants can be overwintered in the open field. For exceptional cold seasons, it is necessary to cover the stock plants with soil or other materials. In cold temperate zone, the selected stock plants require to be moved for overwinter storage such as, cold bed, small tunnels, temporary transplanting, pit storage and others.

Heeling in the Cold Bed or Small Tunnels: The selected stock plants are dug up together with root. The dry and infected leaves are removed. The stock plants are temporarily planted in the cold bed or small tunnels and irrigated. The plants should be covered with straw mat or reed mat when the temperature is lower than 0°C at night. The mats should be rolled up during the day time for ventilation in order to keep the temperature at 0°C at night and 15°C in the day time in the cold bed and small tunnels.

Trench Storage: The trenches with 80-90 cm in depth and 100 cm in width are dug up in the locality of higher terrain and easy drainage. The stock plants are dried in the sun for 3-4 days after harvest and stored in the trench before the ground freezes. The stock plants are covered with soil or other materials gradually along with decrease of air temperature. The thickness of covering material depends on the temperature difference in various regions. The principle of storage is to protect the stock plants against both cold damage and decay because of very thin or very thick covers.

Storage in Vegetable Cellar: The harvested stock plants are stripped off their dry and infected leaves. The stock plants are dried in the sunlight for 3-4 days after harvest and stored in the vegetable cellar before the ground freezes. For efficient use of the space in the cellar, a kind of shelf made of wood stick or bamboo can be made in the cellar. The stock plants are put on each layer of the shelf for storage. The temperature is kept at about 0°C and the humidity at 80-90% within the vegetable cellar.

Management of Stock Plant Transplanting

Basic stock plants of self-incompatible lines are mostly propagated under protected conditions such as cold bed and small tunnels. This method has the following advantages: it is easier to be isolated from other cabbage seed production plots; stock plants can flower earlier; it is easier to find labor to do the hand pollination before busy season. Under protected conditions, stock plants are transplanted in the period of February. The corridor should be left for hand pollination when stock plants are transplanted at a plant space of 30 cm and row space of 30-40 cm. The temperature should be kept at 5-10°C at night and about 15°C in the daytime from transplanting to bolting and budding. During this period, controlled irrigation and inter-tilling should be adopted in order to increase soil temperature and stimulate the growth of root system. In the bolting, flowering and pollination stages, fertilizer and irrigation must be adequately supplied. Ventilation in greenhouse and small tunnels is required to keep the temperature at about 10°C at night and 25°C in the daytime. At the same time, some measures must be taken to control cabbage worm and aphids. Greenhouse or mini-tunnels have to be covered with insect-preventing net to avoid contamination.

At the bolting and flowering stage, the selection of stock plants has to be carried out according to the flowering characters in order to ensure seed purity.

Bud Pollination

Care must be taken not to damage stigmas during stripping. It is difficult to obtain seeds of self-incompatible lines by self-pollination and sister pollination at the flowering stage. Their basic seeds can be propagated by hand pollination at bud stage (Pearson, 1931). The procedure is as follows: The bud top is removed by tweezers and strippers to expose stigma which is then pollinated with pollen collected from the same line. It requires specially trained persons to do the pollination carefully to avoid contamination. Tweezers and hands have to be sterilized with alcohol when pollination at one line is finished and changed to another. The seed production plot must be covered with net to avoid contamination by bees or other insects. Pollen must be freshly collected from the opened flowers on the same day or the day before. If the pollen is collected two days before pollination the seed setting will be enormously reduced. The mixed pollen collected from the same line should be used for pollination to avoid viability depression from continuous selfing. If the bagging isolation is applied to propagate the basic seeds, the flowering branch has to be covered with paraffin bag before the bud opens. When the flowers of the lower inflorescence bloom, bud pollination should be started by using the mixed pollen collected from the same plant under covered bag. After pollination, the flowering branch has to be covered with paraffin bags immediately. The flower buds need to be stripped tenderly and carefully without twisting flower stalk. The bud size neither too small nor too big should be selected for pollination. Bud pollination conducted 2-4 days prior to flowering can give the highest seed set. According to the bud position on the flowering branch, the bud pollination carried out between 5th-20th flower buds above the blooming flower on the inflorescence can be the best for seed setting. But for less vigorous plants, the bud pollination made between

5th-15th flower buds above the blooming flower can produce more seeds.

Propagation of basic seeds of incompatible lines by bud pollination is labor-intensive and costly. In consideration of this disadvantage, the electricity-aided pollination (Roggen and Van Dijk, 1973), wire brush pollination, thermal-aided pollination (Roggen and Van Dijk, 1976), CO_2 enrichment (Nikanishi, Esahi and Hinata, 1975), etc., have been suggested. However, each one of them has its limitations and has not yet been used on commercial scale. In recent years, spraying a solution of 5% common salt has been used to overcome the self-incompatibility and increase the seed set by scientists in China. This method has succeeded in the propagation of basic seeds.

Harvesting the Basic Seed

Harvesting will start 70-80 days after pollination when the seed pods are getting yellow. If the seeds are not mature and are harvested too early, their germination is affected. If it is too late, the seeds will be lost due to pod split. The seeds of some self-incompatible lines have the tendency to germinate within the pod, hence they should be harvested at an earlier stage. After drying, the basic seeds should be maintained in desiccator or in fully sealed container to ensure high germination rate. Seed harvesting, drying and conservation should be managed by specially trained persons. During the whole process, mechanical mixture should be avoided.

Production of F_1 Hybrid Seed

The F_1 Hybrid seed production using self-incompatible approach involves seedling cultivation at fall season, selection of stock, overwintering of stock plants, flowering in the following spring, cross-pollination, seed set, and harvesting of hybrid seed. The whole process takes about 10 months. Each step should be managed properly to obtain high yield and quality of seeds.

Seedling Cultivation for Stock Plants

Sowing: As cabbage requires vernalization, the seedlings of a certain size need to be vernalized. Because parental lines for F_1 hybrid seed production are selected from continuous selfings, their tolerance to stress is relatively poor. The relatively late sowing helps to reduce disease infection at seedling stage, makes stock plants produce less tight heads favoring storage during winter and vigorous growth in the coming year. However, if it is sown too late and the seedlings grow too small, it affects elimination of off types and bad types. The worst leads to a large number of plants not getting vernalized causing abnormal flowering and seed-setting. The most suitable sowing time is to keep stock plants at semi-head stage before overwintering. Parental lines with middle and late maturity would be sown from late July to early August. For early maturity parent lines, it is better to do sowing from middle to late August. Before sowing, the plots for seed beds should have fertile soil with facilities for easy irrigation and drainage. The previous plots used for cabbage seed production should not be chosen as seed beds in order to avoid mixture of remaining seeds from previous crop with parental lines. A level seedbed that is 1.0-1.3 m wide and 7-10 m long is formed. Between beds, a ditch measuring 30-40 cm wide and 10-15 cm deep is dug and linked with field drainage canal. To prevent seedbed against strong sunshine and heavy shower, shading booth should be established over seedbed. The shading booth is covered with reed mat or reed curtain or shading plastic film. If there are showers, it is necessary to add plastic film on the top of booth and protect seedlings from rain damage. The lower edge of shading booth should be at least 30 cm from seedbed in order to have fresh air flow and prevent high temperature during days with strong sunshine. The reed mat or curtain should be covered or removed, in proper time. On clear days, it should be covered at 9 a.m. and removed at 5 p.m. During cloudy days, it is not necessary to cover. When all seedlings have emerged, the booth should be removed. If the soil is rich in nutrients, it is not necessary to apply fertilizer on seedbed. If fertilizer is needed, it should be mixed completely. Before sowing seeds, the seedbed should be fully irrigated. A thin layer of soil will be added on seedbed after irrigation water seeps into the soil. Sowing seeds in a volume of 2-3 g/m^2 is recommended and a layer of 0.5-1 cm depth is covered with sieved soil.

Transplanting: Before field planting, transplanting is needed to get uniform seedlings. The transplanting bed should be prepared similar to that of seedbed. When seedlings grow up to 2-3 leaves, transplanting can be done with a spacing of 10×10 cm^2 and preferably on a cloudy day or late afternoon. Watering should be done immediately after transplanting. Bed should be shaded for 3-5 days and then irrigated once for seedling establishment. Inter-tilling and hardening are required, and spraying are important to control aphids and cabbage worm at the seedling stage.

Field Planting: The field plots for fixed planting should have rich soil, easy irrigation and drainage. If the previous crop was not a crucifer crop, it could reduce disease damage. Base fertilizer should be applied in the field plots in advance and the surface levelled after the till. Field planting will be undertaken when seedlings reach 6-7 leaves. Seedlings with roots in a lump of soil will lead to quick recovery after planting. The density for parental lines with middle or late maturity is about 35×45 cm^2 and for early parental lines 27×33 cm^2. It is better to plant on a cloudy day or late afternoon to protect from strong radiation. Planting and watering should be done at the same time. After seedling establishment, inter-tilling and hardening are done at proper time. The field management for hybrid seed production will not be the same as that of cabbage production in autumn that requires much more watering and fertilizer to produce big cabbage heads. Control of viruses, black rot, downy mildew, etc., is very essential. Much more attention should be paid to spraying chemicals to control cabbage caterpillars, aphids, flea beetle and cabbage moth. Before winter, a less tight head should be formed, otherwise the stock plants are too small and will affect the bolting, flowering and seed-setting in the coming spring. When autumn temperature is low, more careful management should be taken to obtain less tight heads.

Stock Selection

It involves eradication of off-types and bad plants in the field. The off-type plants which have no characteristics of parents may come from biological mixture during reproduction of basic seeds or physical mixture during harvesting and drying in the sun. The bad-type plants refer to diseased, weak and damaged plants. Elimination of off-types and bad-types can be done when seedlings are transplanted. When seedlings grow up and the canopy of seedlings has not yet completely covered plots, extraordinary big or small plants and diseased plants are removed. When the plants reach lotus flower shape, inspection should be done. All plants that have no characteristics of parents should be eliminated firstly based on leaf shape and distribution of leaf vein, and secondly on leaf color and waxy appearance on leaf surface. Before over-wintering, stock plants have formed small head. It is proper time to discard the off-types and bad-types because stock plants have showed characteristics of parental lines under normal cultivation management. All eradication should be done before frost occurs. Otherwise frost-damaged plants show abnormal leaf shape and color, which influence stock selection. When plants start to flower, final rejection should be done based on performance of leaf color, leaf shape, plant vigor, branching, bud shape, bud color, petal color, etc., comparing with parental characters.

Overwintering Management of Stock Plants

How to manage the stock plants of cabbage for overwintering depends on regions with different climates. Plants can overwinter in the open field of warm temperate regions where the lowest temperature is not less than $-10°C$ while in the cold temperate regions the protection facilities are required to help cabbage stock overwinter.

Overwinter in the Open Field: Normally cabbage stock plants have strong tolerance to cold and can even survive -6 to $-8°C$ for a short time after low temperature hardening. However, the plants get damaged if low temperature persists for a longer time. To prevent the stock plants from cold damage during overwintering the following measure can be taken.

Ridge Culture: The stock plants are grown on the slope of ridge from east to west so that ridge can be used as wind proof. By this way plants are placed in warm and enough sunshine microenvironments which help stock in safe overwintering.

Proper Watering: During overwintering cabbage stocks still grow slowly. Before severe cold winter, watering not only means the water demands of plants but also avoid sharp decreases of soil temperature.

Proper Hilling-Up: After watering in the winter it is important to intertill in order to keep soil moisture and hill-up the stem base of stock plants. The proper hill-up height can be as high as 1/2 or 1/3 of plant height.

Overwintering in Storage: There are several methods for cabbage overwintering in the protected areas such as small tunnel, cold bed, storage hidden in the earth, pit storage, etc.

Mini-Tunnel: The parental line with round head type would be grown in the small plastic tunnel for overwintering. Because this type of parents need strong sunshine during overwintering, they can bolt and flower well in the coming year. The tunnel is covered with plastic film and added with straw mat at night which is removed during the day time. Stock plants can receive enough sunshine and grow at the proper temperature. The intensive management should follow according to weather change during winter. In the tunnel, high temperature and spindling should be prevented. Tight head formation will not be good for bolting and flowering in the coming year. The stock plants are planted into small plastic tunnel when freeze appeared in the open field. After planting watering is needed. Several days later, inter-tilling should be done to enhance growth of root system. Irrigation is done according to soil condition in order to meet water demands of plants.

Cold Bed: Before overwintering, stock plants are transplanted into cold bed. Watering is necessary before severe winter comes and also the cold bed is required to be covered with straw mat or reed curtain at night and removed in the day time to receive sunshine. In the early spring when the weather is getting warm, more attention should be paid to prevent stock plants from heat. All mats on the bed must be removed to dry in the sun at the daytime. When the bed is covered with straw mat at night a small outlet should be left. Otherwise, plants are hit with heat to lead to leaves dropping down and affect seed yield.

Storage Hidden in the Earth: This method is suitable for parents with flat round head and sharp round head types. Storage hidden in the earth is to use soil environments with rather stable temperature and moisture. In this way, cold damage to stock plants can be prevented for safe overwintering. Normally, a higher place is chosen to dig a ditch on the shadow side. The ditch width is one meter which should be under the frozen layer and its depth depends on local climate. All stock plants are placed in the ditch with tops of their heads in a position between frozen layer and unfrozen soil. Spacing among plants keeps to avoid heat damage to stock plants. As temperature goes down, more and more soil is covered to protect the stem base and root system from freezing damage. If the ditch is too long or too wide, it is important to put a bundle of corn stray in the middle at a certain interval space in order to be well ventilated and reduce temperature and moisture in the ditch. To meet the seedling size for vernalization, the stock plants to be buried should have small loose head. According to weather changes covered soil is added or removed to allow stock plants safely overwinter.

Pit Storage: For this type of storage, all stock plants are put in the semi-underground tunnel with skylight. Inside tunnel there are 3-4 shelves, 50 cm wide and 150 cm high made of bamboo or wood sticks. Between shelves there is 0.8-1.0 m space for corridor. In each self, 3-5 layers of stock plants are stacked to be well ventilated and rearranged. All stock plants are harvested at 15 days before soil gets frozen and dried in the sun for 2-3 days. All old and infected leaves on the stock are removed. The stock plants are stacked in a way of putting lower parts with roots toward the center and head parts outwards. When the temperature falls below 0°C, the stock plants are moved into the tunnel and rearranged within each other for 15 days. Attention should be paid to ventilation and temperature control in the tunnel during storage. Stock plants for pit storage are required to be as big as that for buried storage. According to weather change, temperature should be controlled by ventilation, and ensuring freeze protection. Furthermore, moisture in the tunnel should be monitored to prevent drying of the stock roots. Otherwise, re-rooting will be affected in the coming year.

Transplanting Methods and Management of F_1 Hybrid Production Plots

Transplanting of stock plants can be done either in the open field or in the protected area. If the flowering of female and male parents is synchronous, it is better to produce F_1 hybrids in the open field. If not, F_1 hybrids should be produced in the protected cultivation such as cold bed, and mini-tunnel. The plant ratio of female to male is 1:1 and planted in alternate rows. If the plant vigor of both parents differ from each other greatly, ratio might be 2:2 which is in favor of insect pollination. For the parents which have different seed set, it is recommended that parent with better seed set should be planted more with maximun ratio of 2:1. For F_1 seed collection, row spacing is 50 cm for inter-rows and plant spacing is 33-37 cm. In the warm temperate region where F_1 hybrid are produced in the open field, normally stock plants are planted in the field before winter comes. After overwinter it is not necessary to plant them again. In the early spring before stock plants turn green, old and infected leaves are removed. Proper watering and inter-tilling with fertilizer application are required to enhance early bolting and flowering. In the case of stock plants which are put in the cold bed, buried storage and pit storage, 60 t/ha of base manure should be applied before stock plants are planted in the field and plots of 10 m length with 1 m width are prepared after leveling the field. For ridge cultivation, ridge width is about 50 cm. Although the planting plots are prepared differently in dofferent regions of China, rational density should be considered. For a combination of mid and late maturity, 60,000 plants/ha is recommended. For early combination, 67,500 plants/ha is suggested. Several aspects for planting stock in the early spring are: (1) plants in cold bed are transplanted with soil which can protect injury of roots, (2) if the soil in the plot has proper moisture, drip irrigation can be adopted to maintain soil temperature and quick rooting of stock plants, and (3) plastic mulch at 10-15 days before planting is suggested to increase soil temperature to help quick rooting.

If the F_1 hybrids are produced in the cold bed or mini-tunnel, stock plants should be planted 15 days before soil gets frozen. Generally 1.233 m width of cold bed for 4 rows and 1.66 m for 5 rows of plants are planted in a space 30-40 cm between plants. For the combination of parents flowering at the same time, the female and male are planted in the direction of south to north at a ratio of 1:1 or 2:2. If the two parents flower at different times, plants are grown in the direction of east to west; late flowering parent on the north of cold bed and early flowering parent on the south so that flowering time of both parents can be adjusted. The plant ratio is still 1:1 or 2:2. Before winter comes, watering and inter-tilling are necessary to keep moisture. In the coming spring before the stock plants turn green, watering can increase soil temperature and help plants to recover. Compound fertilizer can be applied before watering. At the time when stock plants start to bolt, aphid is a major problem and should be well controlled. When the stock plants are in early flowering, sufficient watering and fertilizer should be applied in order to meet the nutrition and water demands of bolting and flowering.

Special Considerations for F_1 Hybrid Production Plots

Isolation Should Be Strictly Guaranteed: Self-incompatible lines of cabbage are easily pollinated by other *Brassica* lines. Therefore, hybrid production plot should be isolated by at least 2000 meters where no cauliflower, kohlrabi, broccoli, kale, Brussel sprouts, etc., are planted. Especially early bolting plants of early cabbage, cauliflower and kohlrabi in the commercial production plot around F_1 hybrid seed production plots should be prohibited.

Bee Population Should Be Large Enough: To improve seed set it is necessary to have sufficient bee population for pollination. Based on experience, at least 15 boxes of bee must be placed in one hectare area for pollination which is helpful to increase hybrid ration and seed yield.

Build Up Framework to Prevent Lodging: The cabbage stock plants get easily lodged during flowering and podsetting. The lodging affects insect pollination and also causes pod mildew, which decreases seed yield and germination. Therefore, at early flowering stage, a framework should be built up with bamboo sticks and straw rope or stick stands beside each plant in order to prevent lodging.

Insect and Disease Control: Major pests are aphids, cabbage worm; main diseases are black rot, viruses, damping-off at seedling stage, and powdery mildew at flowering stage, etc. The control measures include spraying chemicals besides using resistant parents and avoiding continuous cropping with cabbage crops.

Adjustment of Both Parents' Flowering Time: If flowering time of both the parents does not coincide then adjustment of both parents' flowering time is necessary to ensure seed yield and quality. The most important issue is to select inbred lines with similar flowering characteristics as parents. If not possible, overlapped flowering time of both parents can be obtained by adjusting the sowing time, transplanting time and stock plant trimming which leads to synchronization of flowering.

Harvest of F_1 Seed

If parents are self-incompatible lines, harvested seeds from both parents can be mixed up, but to improve the seed uniformity, it is recommended that F_1 seeds of both the parents be harvested separately. When the pod gets waxy yellow and seeds at lower branches of plants are becoming reddish brown it is time to harvest. Plants are dried in time after harvest to prevent seeds from getting infected with mildew. If seeds are harvested prematurely and plants are stacked up too long after harvest, their germination will be adversely affected. Washing of seeds is not allowed if the cleanliness of newly harvested seeds is not good enough, otherwise, solute leakage from seed will reduce germination. After drying and cleaning all new seeds will be stored in a dry and cool (0-40°C) room.

REFERENCES

Bannerrot, H., Loulidard, L., and Tempe, T. (1974). Cytoplasmic male sterility transfer from *Raphanus* to *Brassica*. In Eucarpia-Cruciferae Conference, pp. 52-54.

Borchers, B.A. (1966). Characteristics of a male sterile mutant in purple cauliflower (*B. oleracea* L.). Proceedings of American Society for Horticultural Science.

Cole, K. (1959). Inheritance male sterility in green sprouting broccoli (*B. oleracea* L. var. *italica*). Journal of American Society for Horticultural Science. 95(1): 13-14.

Dickson, M.H. (1997). A temperature sensitive gene in broccoli. Can. J. Genet. Cytol. 1: 203.

Earle, E., Stephenson, C., Walters, T., and Dickson, M. (1994). Cold-tolerant *Ogura* CMS *Brassica*. Vegetables for Horticultural Use. 16: 80-81.

Fang, Zh.Y., Sun, P.T., Liu, Y.M., Yang, L.M., Wang, X.W. and Zhuang, M. (1997). A male sterile line with dominant gene (Ms) in cabbage and its utilization for hybrid seed production. Euphytica, 97:265-268.

Fang, Zh.Y. (1993). Hybrid breeding in cabbage. In Hybrid Breeding in Cruciferae Vegetables. Agriculture Press, pp. 215-275.

Fang, Zh.Y., Sun, P.T., Liu, Y.M. (1991). Cultivation Technology of Cabbage. Jindun Press.

Fang, Zh.Y., Sun, P.T., and Liu, Y.M. (1983). Utilization of heterosis and some problems encountered in self-incompatible line selection in cabbage. Chinese Agricultural Science, 3; 51-61.

Len, H.N. (1988). Main obstacles in the utilization of male sterility in cabbage. Chinese Vegetables, 4: 8-10.

Nasrallah, M.E. and Wallace, D.H. (1968). The influence of modifier genes on the intensity and stability of self-compatibility in cabbage. Euphytica, 17: 493.

Nieuwhof, M. (1961). Male sterility in cole crops. Euphytica, 10: 351-356.

Nikanishi, T., Esashi, T., and Hinata, K. (1975). Self seed production by CO_2 gas treatment in self-incompatible cabbage. Euphytica, 24:117.

Pearson, O.H. (1972). Cytoplasmically inherited male sterility characters and flavor components from the species *Brassica nigra* (L.) *Koch* × *B. oleracea* L. J. Am. Soci. Horti. Sci., 97: 397-402.

Pearson, O.H. (1931). The influence of inbreeding upon the season of maturity in cabbage. Proc. Am. Soc. Hortic. Sci. 29: 359.

Pelletier, G., Primard, C., Vedel, F., Chetrit, P., Remy, R., Rousselle, P., and Renard, M. (1983). Intergeneric cytoplasmic hybridization in crucifereae by protoplast fusion. Mol. Gen. Genet. 191:244-250.

Roggen, H.P., and Van Dijk, A.J. (1976). Thermally aided pollination: a new method of breaking self-incompatibility in *Brassica olenwea* L. Euphytica, 25: 643.

Roggen, H.P. and Van Dijk, A.J. (1973). Electric aided and bud pollination, which method to use for self-seeds on cole crops (*Brassica oleracea* L.). Euphytica, 22: 260.

Sampson, D.R. (1966). Genetic analysis of *Brassica oleracea* using genes from sprouting. Can. J. Genet. Cytol. 8:404-413.

Suteki, Shinohera, (1984). Vegetables Seed Production Technology of Japan. Tokyo YOHKENDOH Co., Ltd.

Index

T - #0592 - 101024 - C0 - 229/152/8 - PB - 9781560220756 - Gloss Lamination